Crystallography

Edited by Takashiro Akitsu

Published in London, United Kingdom

IntechOpen

Supporting open minds since 2005

Crystallography
http://dx.doi.org/10.5772/intechopen.78499
Edited by Takashiro Akitsu

Contributors
Yoshitaka Matsukawa, Arthur Dyshekov, Yurii Khapachev, Takashiro Akitsu

Notice
Statements and opinions expressed in the chapters are these of the individual contributors and not
necessarily those of the editors or publisher. No responsibility is accepted for the accuracy of
information contained in the published chapters. The publisher assumes no responsibility for any
damage or injury to persons or property arising out of the use of any materials, instructions, methods
or ideas contained in the book.

First published in London, United Kingdom, 2019 by IntechOpen
IntechOpen is the global imprint of INTECHOPEN LIMITED, registered in England and Wales,
registration number: 11086078, The Shard, 25th floor, 32 London Bridge Street
London, SE19SG - United Kingdom
Printed in Croatia

British Library Cataloguing-in-Publication Data
A catalogue record for this book is available from the British Library

Additional hard and PDF copies can be obtained from orders@intechopen.com

Crystallography
Edited by Takashiro Akitsu
p. cm.
Print ISBN 978-1-83881-878-4
Online ISBN 978-1-83881-879-1
eBook (PDF) ISBN 978-1-83881-880-7

We are IntechOpen,
the world's leading publisher of
Open Access books
Built by scientists, for scientists

4,200+
Open access books available

116,000+
International authors and editors

125M+
Downloads

Our authors are among the

151
Countries delivered to

Top 1%
most cited scientists

12.2%
Contributors from top 500 universities

CLARIVATE ANALYTICS
BOOK
CITATION
INDEX
INDEXED

WEB OF SCIENCE™

Selection of our books indexed in the Book Citation Index
in Web of Science™ Core Collection (BKCI)

Interested in publishing with us?
Contact book.department@intechopen.com

Numbers displayed above are based on latest data collected.
For more information visit www.intechopen.com

Meet the editor

Takashiro Akitsu, PhD, has been a full Professor for the Department of Chemistry, Faculty of Science Division II, at Tokyo University of Science since 2016.

He completed his undergraduate school training (chemistry) and his graduate school training (physical and inorganic chemistry, especially coordination and crystal and bioinorganic chemistry) at Osaka University in 1995 and 2000, respectively. Dr. Akitsu has published many articles in journals while working at Osaka University (2000–2002), Keio University (2002–2008), and Tokyo University of Science (2008–present). He has been a peer reviewer of many journals and acted on the organizing committees of several international conferences. His research interests are crystal and electronic structures of chiral metal complexes and their hybrid materials.

Contents

Preface

This book reviews a wide range of both current research in and several principles of crystallography, not only for natural sciences, mathematics, physics, chemistry, biology, and earth sciences but also for applied engineering such as material and medical or pharmaceutical sciences. As a review book on crystallography, this book will help with theoretical considerations and understanding the basic theory of frontier experiments, among other topics.

The main themes of this subject can be classified into three categories:

1. Pure mathematical theory or chemical aspects of crystal or molecular symmetry about group theory.

2. Techniques of crystal structure analysis such as experiments on neutron diffraction, computational methods about phase problem, and commonly used crystal structure analysis for chemical compounds.

3. Compounds or topics solved by crystallography such as a review of structural inorganic chemistry and hydrogen bonds in the crystal chemistry of organic compounds.

Each chapter demonstrates all aspects of current crystallographic study.

The reader will also appreciate the many new developments in this subject.

Takashiro Akitsu
Department of Chemistry,
Faculty of Science,
Tokyo University of Science,
Japan

Chapter 1

Introductory Chapter: Crystallography

Takashiro Akitsu

1. Book reviews about "greater" crystallography

Indeed, crystallography consists of wide range of not only natural sciences, mathematics, physics, chemistry, biology, and earth sciences but also applied engineering such as material and medical or pharmaceutical sciences. Like chapters in this book, I have published several book reviews of crystallographic books so far. The themes of these books are roughly classified into three categories:

1. Pure mathematical theory [1] or chemical aspects of crystal [2] or molecular [3] symmetry about group theory.

2. Techniques of crystal structure analysis such as experiments of neutron diffraction [4], computational methods about phase problem [5], and commonly used crystal structure analysis for chemical compounds [6, 7].

3. Compounds or topics solved by crystallography such as a review of structural inorganic chemistry [8] and hydrogen bonds in crystal chemistry of organic compounds [9].

Crystallographic books, of course like this book, may play a helpful with theoretical consideration or comparison with previous examples and so on.

2. Problems in crystallographic study

To data, however, my crystallographic study [10, 11] on single crystal or powder structure analysis of chemical compounds especially (chiral) metal complexes has been suffering things. I have also challenged to investigate hybrid materials composed of metal complexes and other materials such as metal nanoparticles and proteins, which are usually dealt with other types of crystallographic experiments. In other words, chemical and structural-biological (protein) single-crystal analyses are similar to each other in principle, though they are different from the actual. Combination of several techniques of crystallography should be employed or developed for these studies. Probably, spectroscopy and crystallography may be good partner to be used at the same time.

One of the serious problems may be basic level, namely, poor quality of crystal samples as a simple component for desiring hybrid materials. For both single crystal using laboratory MoKα radiation (**Figure 1** up) and powder diffraction even by synchrotron beam (**Figure 1** down), rings due to low resolution or wide peaks sometimes appeared as shown in **Figure 1**.

I want to discuss more essential problems furthermore in order to establish integrated crystallography hybrid materials in the future.

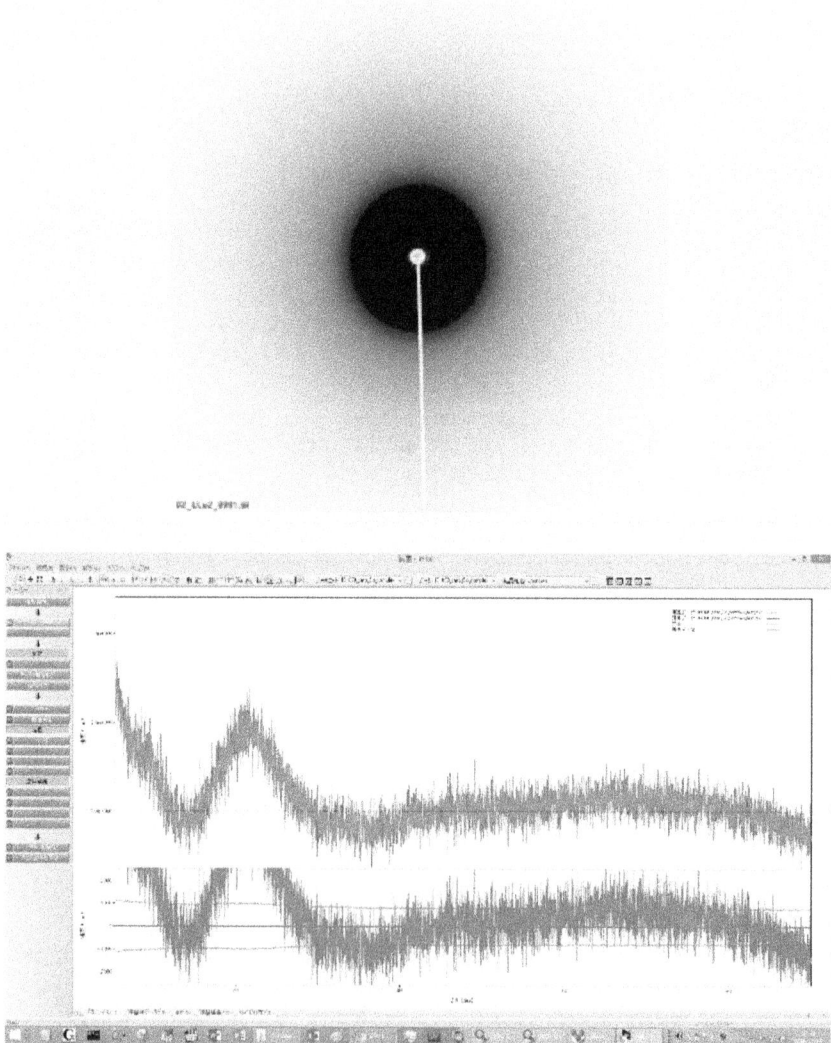

Figure 1.
Bad diffraction patterns of single crystal (up) and powder (down).

Author details

Takashiro Akitsu
Department of Chemistry, Faculty of Science, Tokyo University of Science, Tokyo, Japan

*Address all correspondence to: akitsu2@rs.tus.ac.jp

IntechOpen

References

[1] Akitsu T. Book review: Crystal group–Kyoritsu series of mathematics quest 7. Journal of the Crystallographic Society of Japan. 2016;**58**:284. https://doi.org/10.5940/jcrsj.57.252

[2] Akitsu T. Book review: IUCr monographs on crystallography 17 symmetry in crystallography-Understanding the international tables. Journal of the Crystallographic Society of Japan. 2012;**54**:234. https://doi.org/10.5940/jcrsj.54.234

[3] Akitsu T. Book review: Symmetry of Crystals and Molecules. Journal of the Crystallographic Society of Japan. 2015;**57**:252. https://doi.org/10.5940/jcrsj.57.252

[4] Akitsu T. Book review: Inorganic chemistry based on quantum theory, approach from group theory. Journal of the Crystallographic Society of Japan. 2010;**52**:191. https://doi.org/10.5940/jcrsj.52.Book06

[5] Akitsu T. Book review: Applications of neutron powder diffraction. Journal of the Crystallographic Society of Japan. 2009;**51**:267. https://doi.org/10.5940/jcrsj.51.Book04

[6] Akitsu T. Book review: IUCr monographs on crystallography 20 phasing in crystallography–A modern perspective. Journal of the Crystallographic Society of Japan. 2014;**56**:338. https://doi.org/10.5940/jcrsj.56.338

[7] Akitsu T. Book review: IUCr monographs on crystallography 14 crystal structure analysis–A primer third edition. Journal of the Crystallographic Society of Japan. 2010;**52**:303. https://doi.org/10.5940/jcrsj.52.Book06

[8] Akitsu T. Book review: IUCr monographs on crystallography 14 crystal structure analysis–A primer third edition (Japanese translated version). Journal of the Crystallographic Society of Japan. 2011;**53**:362. https://doi.org/10.5940/jcrsj.53.362

[9] Akitsu T. Book review: IUCr Monographs on Crystallography 23 The Nature of the Hydrogen Bond ~Outline of a Comprehensive Hydrogen Bond Theory~. Journal of the Crystallographic Society of Japan. 2010;**52**:190. https://doi.org/10.5940/jcrsj.52.Book03

[10] Akitsu T. Book review: Advanced Structural Inorganic Chemistry, IUCr Text on Crystallography 10. Coordination Chemistry Reviews 2009;**253**:2782. https://doi.org/10.1016/j.ccr.2009.06.002

[11] Akitsu T, editor. Crystallography: Research, Technology and Applications. NY, USA: Nova Science Publishers, Inc.; 2012. ISBN: 978-1-62081-574-8

Chapter 2

Crystallography of Precipitates in Metals and Alloys: (1) Analysis of Crystallography

Yoshitaka Matsukawa

Abstract

This chapter and the following chapters describe crystallography of second-phase precipitate particles in metals and alloys. The focus of this chapter is placed on technical aspects in the analysis of their crystal structure, composition, and crystal orientation relationship with the matrix. Characterization of fine precipitates embedded in solid matrix is technically rather difficult; the signal from the matrix always hinders the signal from the precipitates. Although even state-of-the-art characterization techniques are still incomplete, it is becoming possible to assess the validity of assumptions involved in classic theories related to the crystallography of precipitates. For instance, recent experimental studies demonstrated that evolution of their crystal structure during nucleation seems to contradict the so-called classical nucleation theory, in terms of fluctuations in size and composition. Recent studies also demonstrated that their crystal orientation relationship with the matrix is often different from the one predicted by energy considerations related to the interfacial lattice mismatch. Furthermore, crystal orientation relationship with the matrix was found to be a factor controlling the magnitude of precipitation hardening, contrary to the conventional Orowan's hardening model based on continuum elasticity theory calculations without considering crystallography.

Keywords: precipitates, nucleation, crystal structure, strength, dislocations

1. Introduction

This chapter and the following chapters review recent progress of our knowledge about crystallography of precipitate particles in metals and alloys [1–3]. The main focus is placed on the following three subjects:

1. Evolution of crystal structure during nucleation

2. Crystal orientation relationship with the matrix

3. Effect of crystallography of precipitates on mechanical properties

These subjects are closely related to the following three basic theories, each of which has a long history greater than a half century:

1. The theory of crystal nucleation (since 1876) [4]

2. The theory of dislocations (since 1934) [5–9]

3. The theory of precipitation hardening (since 1954) [10, 11]

From an engineering viewpoint, the knowledge provided here is primarily useful for developing stronger materials. Dispersing fine precipitate particles over the matrix at high density is a common engineering technique for improving the strength of metals and alloys. By introducing a minor amount of second-phase precipitate particles, such as 2% in volume fraction, the material strength is increased by several times greater. In the traditional theory of precipitation hardening (a.k.a. dispersion strengthening) established in the 1950s–1960s, the primary factor controlling the magnitude of strengthening effect is assumed to be the shear modulus [10, 11], whether or not precipitates are harder than the matrix. This concept has been partly revised in the past few years. Recent experimental studies using state-of-the-art material characterization techniques demonstrated that crystallography of precipitate particles is another factor dominating their obstacle strength [1, 2]. When the slip plane of dislocations in precipitates is not parallel to that in the matrix, dislocations are unable to cut through the precipitates, resulting in large hardening, regardless of the shear modulus. This subject is extensively discussed in the next chapter.

This chapter may also be of interest for the audience outside of the research community of materials science and solid-state physics. Nucleation is one of the areas of basic science related to a wide variety of research subjects including chemical reactions in liquid and gas. In fact, the first theory was originally developed for the nucleation of droplets from gas. Nucleation of crystals in solid is more complicated than the situation assumed in liquid and gas, in a sense that the formation of a new crystal is highly constrained by the surrounding matrix, in terms of the strain energy associated with the precipitate/matrix interface and the diffusivity of atoms for their agglomeration. A long-standing open question is the critical condition for nucleation regarding size and composition of nucleus. Precipitates are in many cases compounds consisting of multiple elements such as carbides and oxides. Unlike in gas and in liquid, the diffusivity of each element is not the same in solids [12]. For instance, the diffusivity of light elements like carbon and oxygen is several orders of magnitude greater than that of metallic elements. Although the classical nucleation theory assumes that the crystal structure and composition of precipitates are the same as those of the final product from the beginning of embryo growth (**Figure 1**), the diffusivity difference indicates a possibility that the composition of precipitates fluctuates during the nucleation process. The classical nucleation theory also assumes that nucleation occurs when the embryos have grown up to a critical size. In many cases the critical size of precipitates for nucleation is 2–3 nm [3]. Assessing the composition of such small precipitates has been technically impossible until recently. The highlights of recent studies are discoveries that, in the early stage of precipitation, the crystal structure and composition of precipitates are different from those of the final product and that the precipitates structurally transform into the final product at a critical size with a critical composition (**Figure 1**). Precipitates are clusters of solute elements when they start spontaneous growth, which is defined as the state of "nucleation" in the classical nucleation theory. An implication of this finding is that the obstacle strength of precipitates in precipitation hardening may change during precipitation. They are weak obstacles in the early stage of precipitation regardless of the crystal structure of the final product. They can become strong obstacles due to a change in the shear modulus or

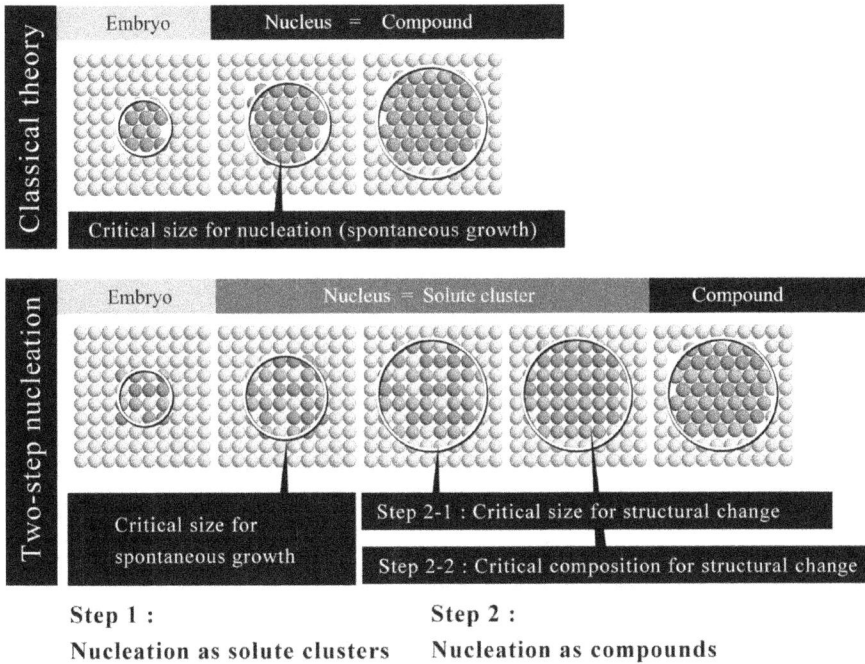

Figure 1
Nucleation of precipitates in metals and alloys: classical nucleation and two-step nucleation [3]. Unlike the classical nucleation theory, in reality, crystal nuclei do not emerge directly from the matrix. They first nucleate as solute clusters structurally indistinguishable from the matrix, followed by a structural change. Their crystal structure changes at a critical size with a critical composition.

the crystal structure. In some cases, precipitates become brittle by the structural change, while they are ductile in the state of solute clusters. Brittle precipitates are considered to serve as the nucleation site of cracks via particle cracking. Hence, from the viewpoint of fracture mechanics, the ductile-brittle transition of precipitates during precipitation considered a factor controlling the engineering lifetime of materials.

As a result of the constraints from the surrounding matrix, precipitation of the second phase often occurs with a specific crystal orientation relationship with the matrix. Precipitates and matrix share a specific atomic plane in such a way to minimize the mismatch between them. The orientation relationship is dependent on their crystal structure. For instance, in the Burgers orientation relationship, bcc precipitates in hcp matrix share atomic planes as follows (**Figure 2**) [13]: $(0001)_{hcp}//(110)_{bcc} \wedge (2\bar{1}\bar{1}0)_{hcp}//(1\bar{1}1)_{bcc}$. Since the lattice parameter is specific to materials, a preferable orientation relationship changes depending on the degree of mismatch of lattice parameter between precipitates and matrix. The Burgers orientation relationship is the optimum configuration for the combination of bcc pure Zr and hcp pure Zr, but another orientation relationship is preferred for the bcc Nb precipitates containing a few amount of Zr. The Zr-Nb binary system is a complete solid solution in a bcc structure at high temperatures [14]. The difference of lattice parameter between the bcc Zr and the bcc Nb is ~10% [15]; the lattice parameter of bcc precipitates changes in accordance with Vegard's law [2]. Apart from a remarkable progress in theoretical works on the orientation relationships, experimental studies have recently demonstrated that precipitates and matrix do not always follow such a theoretically predictable, ideal orientation relationships in reality. Recent analysis using electron backscatter diffraction (EBSD) (**Figure 3**)

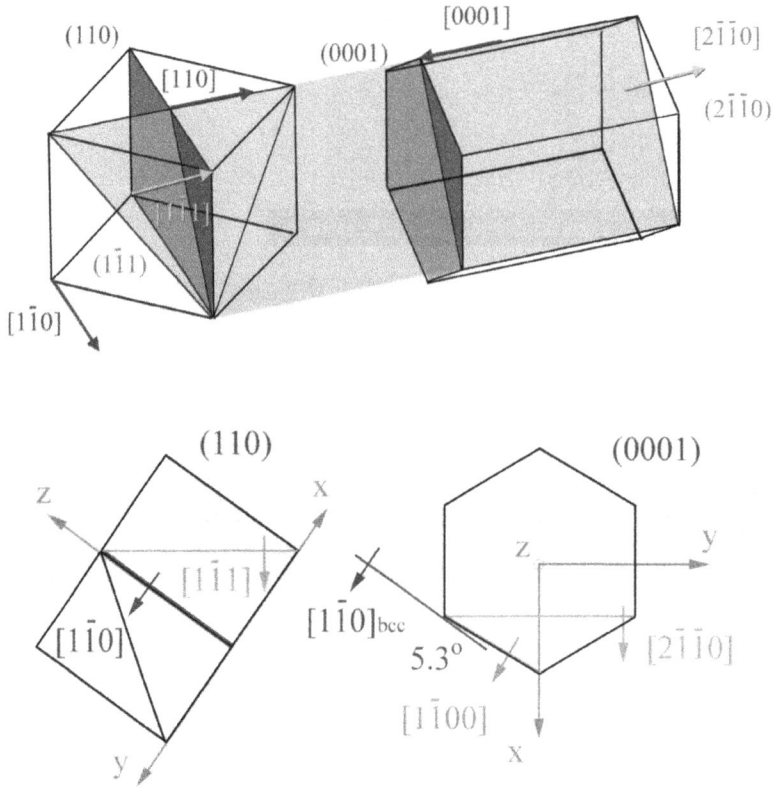

Figure 2
The Burgers orientation relationship for bcc and hcp crystals [2]. This is the most traditional orientation relationship discovered in 1934.

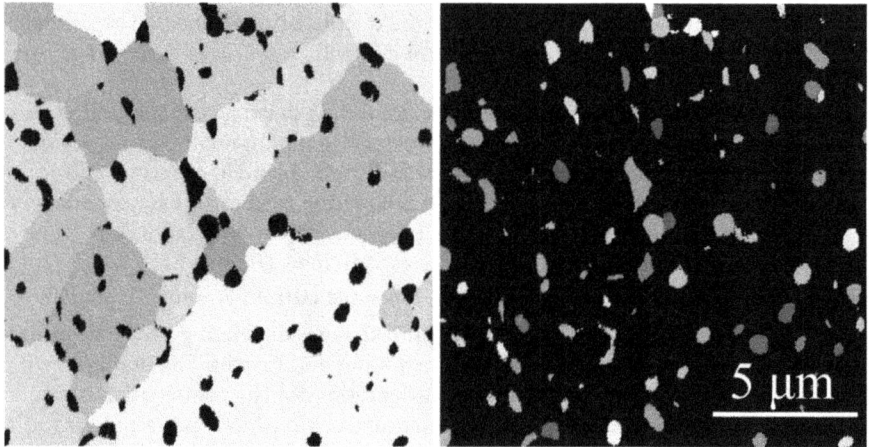

Figure 3
Example of EBSD analysis of precipitates: bcc Zr precipitates containing Nb and hcp Zr matrix in a Zr–2.5Nb alloy [2].

revealed that, when the matrix undergoes recrystallization after precipitation of precipitates, their orientation relationship is overwritten. As a result of that, crystal orientation of precipitates can become random (**Figure 4**). The degree of

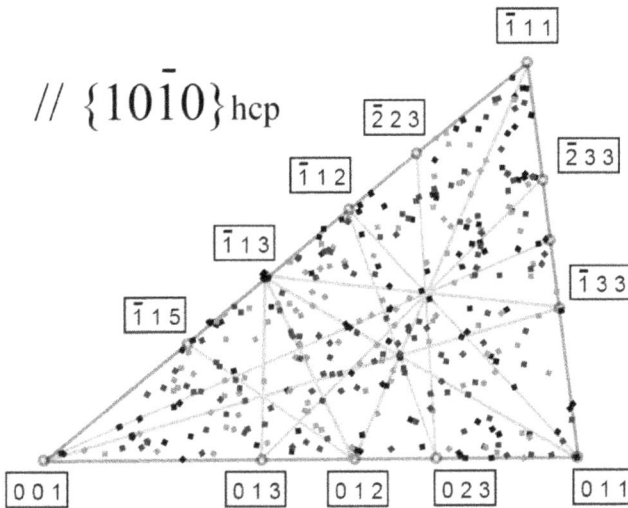

Figure 4
EBSD analysis results of atomic planes of precipitate particles parallel to the slip plane of matrix: bcc Nb precipitates and hcp Zr matrix in a Zr–2.5Nb alloy [2]. Only 1 out of 100 precipitate particles had a slip plane parallel to that of the matrix. Hence, dislocations are unable to cut through the bcc Nb precipitates.

contribution of precipitates to the strength of materials may become different from what is expected from well-known crystal orientation relationships [2].

Recent updates of these theories have been achieved by progress in material characterization methods for determining the crystal structure and composition of nano-sized precipitates. Before going into the details of these theories, we briefly review the technological breakthrough in experimental methods. This chapter is addressed to not only the specialists of precipitates but also nonspecialists including students. For better understanding, traditional methods of material characterization are also briefly reviewed at the beginning.

2. Brief history of microstructure characterization techniques

Crystal structure is determined based on the concept of diffraction, discovered in 1912. It appears that X-ray diffraction (XRD) became common in the 1920s; a great many structures of alloys were determined. Early works determined simple structures having a high symmetry with which peaks in the XRD spectrum are clearly resolved free from overlapping. Precipitates are, however, in many cases compounds having a low symmetry. XRD became applicable to such complicated structures by the invention of the Rietveld method in 1966 [16]. Precipitates involved in bulk metallic samples are detectable only when their volume fraction is higher than ~1% [17], though that is highly dependent on their crystal orientation relationship with the matrix. In bulk samples the crystal orientation of precipitates is not necessarily random, and the matrix grains also not. Metallic bulk samples cannot be crushed into powders due to their high ductility. They can be mechanically grinded into powders by using a hand grinder; however, the XRD peaks of such grinded metallic powders are broadened due to introduction of dislocations, resulting in hindering the peaks of precipitates by the background noise. These issues are avoided by the use of residue extracted from the matrix via chemical dissolution using an acid [18]. This extraction residue analysis is, however, applicable to only nonmetallic compound precipitates embedded in metallic matrix.

Transmission electron microscope (TEM) is a multifunctional characterization tool capable of determining not only crystal structure but also composition and size of precipitates on the image of microstructure, free from the constraint due to volume fraction. The first prototype was produced by Ruska et al. in 1932, and the first commercial model was released by Siemens in 1939. It appears that TEM became common in the 1950s; for example, the number of commercial products released in Japan was greater than 250. The resolution (point resolution) was 50 nm for Ruska's first TEM, 1 nm for the Siemens Elmiskap I released in 1956, and 0.2 nm for the JEOL JEM100B released in 1968. Precipitates are visualized using diffraction contrasts; those satisfying the Bragg condition exhibit dark contrast in the so-called bright-field image (bright contrast in the dark-field image), whereas the others are indistinguishable from the matrix. The number density of precipitates determined by diffraction contrast images represents the true number density only in the case where precipitates are all aligned to the same crystal orientation. This condition is achievable only when precipitation occurs with a specific crystal orientation relationship with the matrix such as the cube-on-cube orientation relationship, where the unit cells of the precipitate and the matrix completely overlap each other. In the other orientation relationships, some crystallographic variants are often invisible. This is a potential error in the evaluation of the number density of precipitate particles but often out of consideration. In many cases, the magnitude of error bars is determined solely by a statistical analysis: either the standard error or standard deviation.

High-resolution (HR)-TEM is another mode capable of visualizing precipitates using phase contrasts, i.e., lattice fringes generated by interference of transmitted and diffracted electron waves. This imaging mode became common in the 1970s–1980s [19]. In those days, however, alignment of electron beam axis was technically difficult for entry-level users. This technical issue was resolved in the 1990s by an introduction of the field-emission gun, which provides a hundred times brighter illumination, a digital camera system, a real-time image processing software (fast Fourier transformation for the alignment minimizing the objective lens stigmatism), etc. However, even though the issue of beam alignment has been resolved, HR-TEM analysis of nano-precipitates is still extremely time-consuming due to alignment of crystal orientation. The HR-TEM image (crystal lattice image) is obtained only when the direction of incident electron beam is aligned with the crystal's zone axis having a low index, e.g. [001] and [110]. The beam-crystal alignment, achieved by using Kikuchi lines or bend counters, is easy for large precipitates greater than several hundred nm but technically almost impossible for nano-precipitates. So for this reason, in practice, the operator searches particles which already exhibit the crystal lattice image without tilting the sample. Unless otherwise precipitates have a specific orientation relationship with the matrix, the operator can find only a few but not many such particles, whereas the minimum requirement of the number of precipitates for drawing a smooth histogram of the size distribution is ~500 in the author's experience [3].

TEM is capable of determining the crystal orientation relationship between precipitates and matrix, though this analysis is also extremely time-consuming. In order to determine the orientation relationship, one needs to find out a sample-tilting angle, where the beam direction is aligned with a zone axis. Three such tilting angles need to be found for both precipitates and matrix in order to determine their (hkl) indices. In some cases precipitates may not have any specific orientation relationship with the matrix; however, proving such a random orientation relationship is practically impossible for one-to-one analysis using a TEM. The random orientation issue can be assessed only if the number density of precipitates is sufficiently high enough for obtaining the Debye ring patterns in selected-area electron diffraction. A more appropriate method rather than TEM to investigate this research subject is EBSD equipped on a scanning electron microscope (SEM). EBSD determines the orientation of crystals based on the Kikuchi pattern, whose theoretical accuracy is ~0.1° [20, 21], whereas the accuracy of orientation analysis using diffraction spots is ~3° [22, 23].

The first report introducing the principle of EBSD was published in 1973, within 10 years after the release of the first commercial SEM, the Stereoscan series 1, by the Cambridge Instrument Company in 1965. EBSD became a practically useful tool in 1993, by full automation of mapping (detecting, indexing, and recording the Kikuchi bands based on the Hough transformation). The spatial resolution of EBSD is dependent on probe size, step size of scanning, accelerating voltage of electrons, sample geometry (bulk or thin foil), etc. According to the author's experience, precipitates of ~500 nm in diameter can be identified but ~50 nm not. The spatial resolution is improved by using an advanced technique called transmission Kikuchi diffraction (TKD), a.k.a. transmission EBSD, proposed in 2012 [24]. This new technique works on conventional EBSD system and software. The difference is that TKD uses forward-scattered electrons, whereas EBSD uses backscatter electrons. In other words, TKD uses transmitted electrons as well as TEM; hence, the samples must be thin foils. Sample preparation is not difficult for TEM users; TEM samples can be directly subjected to this analysis. The high spatial resolution of TKD owes not only to the use of thin foil specimens, which minimize unfavorable lateral beam spreading inside the specimens, but also to a greater signal intensity of forward-scattered electrons than backscattered electrons [25]. Since the Kikuchi pattern is generated from elastic scattering (diffraction) of inelastically scattered electrons [26], there exists a lower limit in both specimen thickness and precipitate size below which the Kikuchi patterns are not obtained. When the thickness of thin foil specimens is largely greater than the size of precipitates, the signal from the precipitates is hindered by that from the matrix. In other words, there exists an upper limit of measurable foil thickness depending on the size of precipitates. Only a limited range of thickness is applicable to this method in a wedged-shaped TEM thin foil specimens. The practical spatial resolution limit of TKD is dependent on many factors such as the position of detector (florescent screen); according to the author's experience using a conventional EBSD system, precipitates of ~50 nm in diameter can be identified but ~10 nm not. The resolution will be improved if the detector is placed just beneath of the sample; this is an ideal setting that minimizes the loss of forward-scattered electrons.

Traditionally, TEM has been a primary analysis tool for composition analysis of precipitates: energy-dispersive X-ray spectroscopy (EDS) and electron energy loss spectroscopy (EELS). In these TEM-based composition analyses, samples having a 3D geometry are projected on 2D space via electron transmission. Precipitates often overlap the matrix in the thickness direction, whereas their TEM image is constructed based on integrated information over thickness. These analyses are unable to determine the composition of overlapped portion. It is practically impossible to judge from the projected 2D image if the precipitates are free from overlapping. In terms of composition analysis of precipitates, the most innovative breakthrough in the past two decades is probably the invention of atom probe tomography (APT). Although its concept was first proposed in 1967, it has become a practically useful tool since the commercial release of local-electrode atom probe (LEAP) in 2003. APT is capable of visualizing atoms in 3D space, which is a critical advantage over the TEM-based composition analyses. APT is a quantitative mass analysis, whereas EDS and EELS are semiquantitative analyses that require a standard sample for calibration. Furthermore, EDS is inherently lack of quantitative accuracy in detection of light elements; emission of Auger electrons is dominant over characteristic X-ray and is dominant for low-Z elements like oxygen. Although APT is superior to TEM-based analyses in many aspects, determination of precipitates' composition is a challenging subject even for APT. The quantitative precision of the APT composition analysis is often limited by artifacts partly due to the so-called trajectory aberration [27–29]. For precipitates darkly imaged in FIM (i.e., low evaporate field regions) compared to the surrounding matrix, defocused high-field iron ions coming from the surrounding matrix fall into the precipitate image on the detector [30]. Conversely, for precipitates brightly imaged in FIM, image overlapping occurs outside the precipitate image. In both cases, mixing with the matrix elements inevitably occurs at the interface. Hence, matrix elements are often detected in nano-precipitates [31, 32].

3. Evolution of crystal structure during nucleation

The classical nucleation theory is based on the so-called capillarity approximation, which assumes that the properties of nuclei are the same as those of the final product from the beginning of embryo growth. In other words, all parameters that characterize the new crystal phase to be distinct from the matrix phase, such as density, composition, and structure, are assumed to be unchanged throughout the nucleation stage. Under this assumption, nucleation event is expected to be solely controlled by the size of embryos. Spontaneous growth (nucleation) of precipitates is expected to occur at a critical composition where the hierarchy of the bulk free energy of the precipitate phase and the surface free energy of precipitate/matrix interface is reversed. In the past two decades, a modern concept called the two-step nucleation has been established by the research community of crystal nucleation

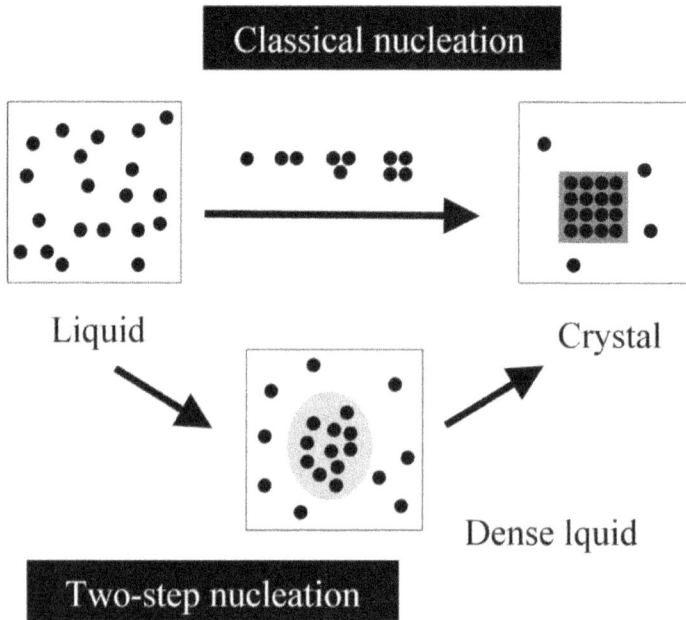

Figure 5
Two-step nucleation of crystals from liquid [33].

from liquid (**Figure 5**) [33]. This concept is nonclassical in that embryos become distinct from the matrix liquid in terms of density prior to the structural change. Here, fluctuation of composition is generally out of consideration, as in liquid diffusivity of solute elements is equally very high; composition fluctuation is expected to be negligibly small. On the other hand, in the nucleation of compound precipitates in crystalline solids, fluctuation of density is relatively small (compared to the nucleation of solid from liquid), but instead, composition may be a variable parameter. In solids, diffusion coefficients of solute elements are merely the same; stoichiometric composition of the compound may not be fulfilled in the early stage of embryo growth. In this case, although the parameter being in focus is different from the conventional two-step nucleation in liquid, this is also nonclassical in the sense that multiple parameters required for nucleation evolve in parallel during nucleation. In 2014, Peng et al. demonstrated that a solid-solid phase transition occurs in a two-step process [34]. In their experiments using a model crystal consisting of microgel colloidal spheres, the two-step represents a two-step change in structure. The first step is a transition from a two-dimensional square lattice structure to a liquid-like structure, and the second step is a transition from the liquid-like structure to a two-dimensional triangular lattice structure. Fluctuation of composition is not associated with their two-step process. Within the framework of the classical nucleation theory, in 1937 Borelius assumed that composition is a variable parameter in the nucleation of precipitates in solids [35]. Absolute value of the bulk free energy of precipitates becomes the greatest with the compound's stoichiometric composition; nucleation is expected to occur at this critical composition. Borelius did not discuss the effect of compositional fluctuation on the critical size. In 1949, Hobstetter attempted to handle both size and composition as variable parameters [36]. He demonstrated that in this two-variable analysis there is a pathway (in terms of evolution of size and composition) energetically more favorable than the pathway fixed by the previous one-variable analyses. However, the meaning of the energetically most favorable pathway remained unclear in the context of critical size and composition.

The final product described in the classical nucleation theory is not necessarily the most stable, equilibrium phase. In many cases, the first nucleating phase is a metastable phase, formation of which occurs with the lowest energy barrier; the equilibrium phase is produced through multiple transitions from a metastable phase to another metastable phase step-by-step. This is an empirical rule known as Ostwald's rule of stages, proposed in the 1890s [37, 38]. One of such examples is precipitation of Al_2Cu at Guinier-Preston (GP) zone in Al-Cu alloys [39, 40]. The precipitation of Al_2Cu, which is the stable phase in this system, is known to occur via multiple intermediate configurations such as GP zone → coherent θ" phase → semi-coherent θ' phase → incoherent θ phase (Al_2Cu). Those intermediate phases are distinct from the Al_2Cu in both crystal structure and composition. Another example is precipitation of fcc Cu in bcc Fe matrix. Precipitation of Cu is known to occur via multiple intermediate configurations such as bcc Cu → a twinned 9R Cu → fcc Cu [41]. The bcc Cu precipitates are crystallographically indistinct from the matrix; in other words, they are solute clusters in the bcc solid solution. The critical composition for their structural changes remains unclear. It is technically rather difficult to determine the composition of precipitates in the early stage of precipitation due to their small sizes.

Traditionally, experimental studies on the nucleation in solids have focused on determining the critical size. For example, Othen et al. [41] reported that the bcc Cu precipitates grow with the twinned 9R structure in a size range from 6 to 15 nm. Their conclusion is based on the results of HR-TEM observation. This methodology is, however, insufficient for statistical argument as mentioned in the previous section.

Even today, experimental studies on the critical composition for nucleation are still limited. As mentioned earlier, even in atom probe tomography, mixing with matrix elements inevitably occurs at the precipitate/matrix interface due to trajectory aberration. Hence, it is practically impossible to judge if the abovementioned Cu precipitates embedded in Fe matrix is 100% pure Cu. When the precipitate of interest is a compound consisting of multiple elements, the ratio of its constituent elements can be discussed. However, when one of those elements is the element of matrix, such as the Al–Cu precipitates in Al alloys, interpretation of their concentration ratio is not straightforward.

In order to determine the critical composition for the structural change, the crystal structure of precipitates must be examined together with composition. In TEM observation of diffraction contrasts, precipitates are indistinguishable from the matrix while they are solute clusters, and they become visible after structural change. By using this unique feature in visibility, recently, Matsukawa et al. performed a systematic analysis on the precipitation of the G-phase in a duplex stainless steel subjected to thermal aging [3]. The crystal structure of the G-phase is cF116 (a variant of fcc structure), and the lattice parameter is exactly fourfold of the matrix ferrite (**Figure 6**). Precipitation occurs with the cube-on-cube orientation relationship [42]. The stoichiometric composition is $Ni_{16}Si_7Mn_6$; its constituent elements are different from the matrix elements (Fe and Cr). So for these reasons, this intermetallic compound is ideal for the fundamental study of nucleation. Precipitation of G-phase in duplex stainless steels is known to occur only in a very narrow temperature range, 673–773 K [43]. In their study, thermal aging was performed at 673 K for up to 10,000 h.

Their analysis revealed that precipitation of Ni–Si–Mn clusters started at 500 h (**Figure 7**), whereas their structural change transforming into the G-phase started at 10,000 h (**Figure 8**). The number density of G-phase particles detected by TEM was only ~26% of the number of Ni–Si–Mn precipitates detected by APT. In other words, three quarters of the Ni–Si–Mn precipitates were solute clusters yet without structural change. The number of particles examined by TEM was ~750. A potential error factor that could cause a misevaluation of the precipitate number density is the method used to evaluate the thickness of the TEM foil. Their method was to use thickness fringes obtained at an exact Bragg condition, where the deviation parameter was s=0. In this case, thickness is determined by the number of thickness fringes multiplied by the extinction distance of the electron beam. Since the precipitate number density was counted in portions where the number of thickness fringes was 4, the magnitude of the error in the foil-thickness evaluation was ±25%. In other words, the number of Ni–Si–Mn clusters that exhibited the crystal structure change was at most 50% of the total.

Their APT analysis also revealed that the Ni–Si–Mn clusters contained not only the G-phase elements (Ni, Si, and Mn) but also the matrix elements (Fe and Cr) and that enrichment of the G-phase elements occurred during thermal aging. Unlike the size growth, the solute enrichment continued even after 5,000 h. In the composition analysis of the clusters (**Figure 9**), those clusters were divided into three groups by size, i.e., small (<2 nm in diameter), medium (2–3 nm), and large (>3 nm), in order to minimize the artifacts that occur at the cluster/matrix interface; a comparison of cluster composition should be made for those having the same size. The concentration ratio of the G-phase elements (Si/Ni and Mn/Ni) did not change during the isothermal aging. The Mn/Ni ratio was in good agreement with that of stoichiometric composition, whereas the Si/Ni ratio was roughly a half of the stoichiometric ratio.

Their analysis indicates that the nucleation of the G-phase occurred via a two-step process: the first step is the spontaneous growth of solute clusters (i.e., nucleation as solute clusters), and the second step is the nucleation as compounds (i.e., the G-phase) (**Figure 1**). There was a time lag between the end of size growth

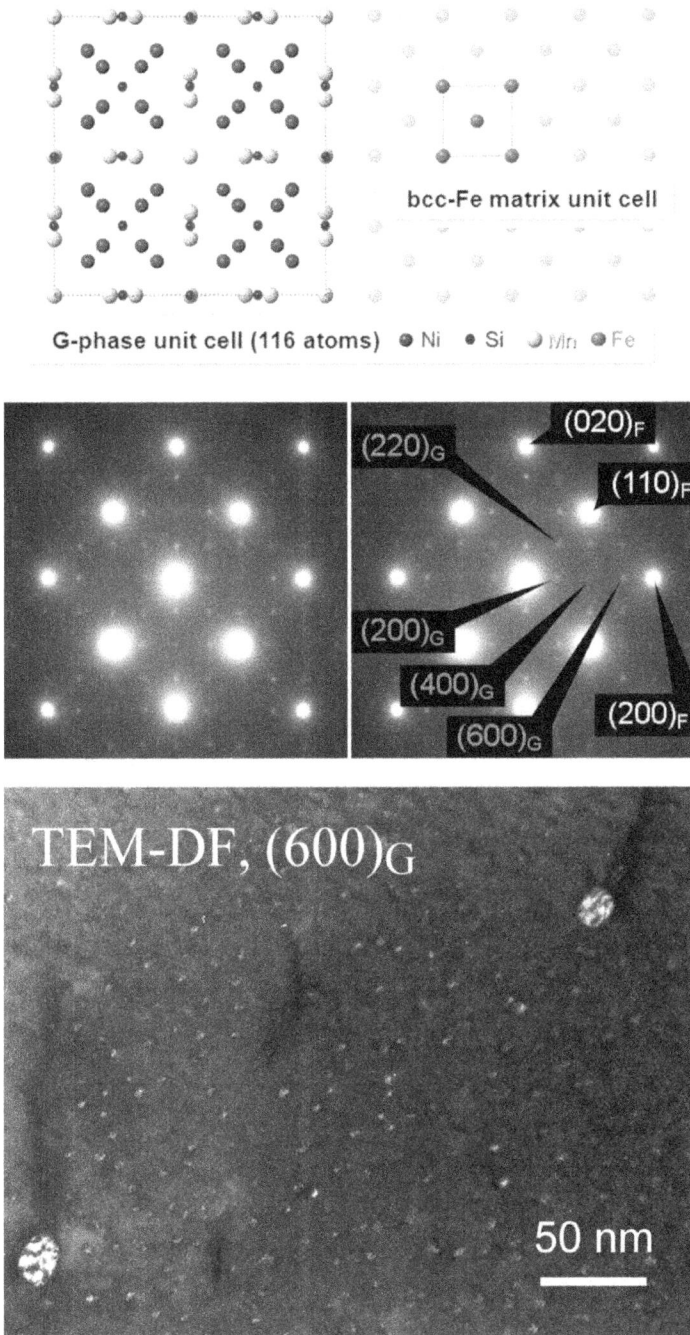

Figure 6
Crystal structure and TEM electron pattern of the G-phase precipitates in the ferrite portion of a duplex stainless steel subjected to thermal annealing at 673 K [3].

(5,000 h) and the start of structural change (10,000 h). It appears that the incubation period was controlled by solute enrichment inside the clusters. In other words, the structural change occurred via another two-step process: the first step is size fluctuation to become a critical size, and the second step is composition fluctuation to become a critical composition (**Figure 1**).

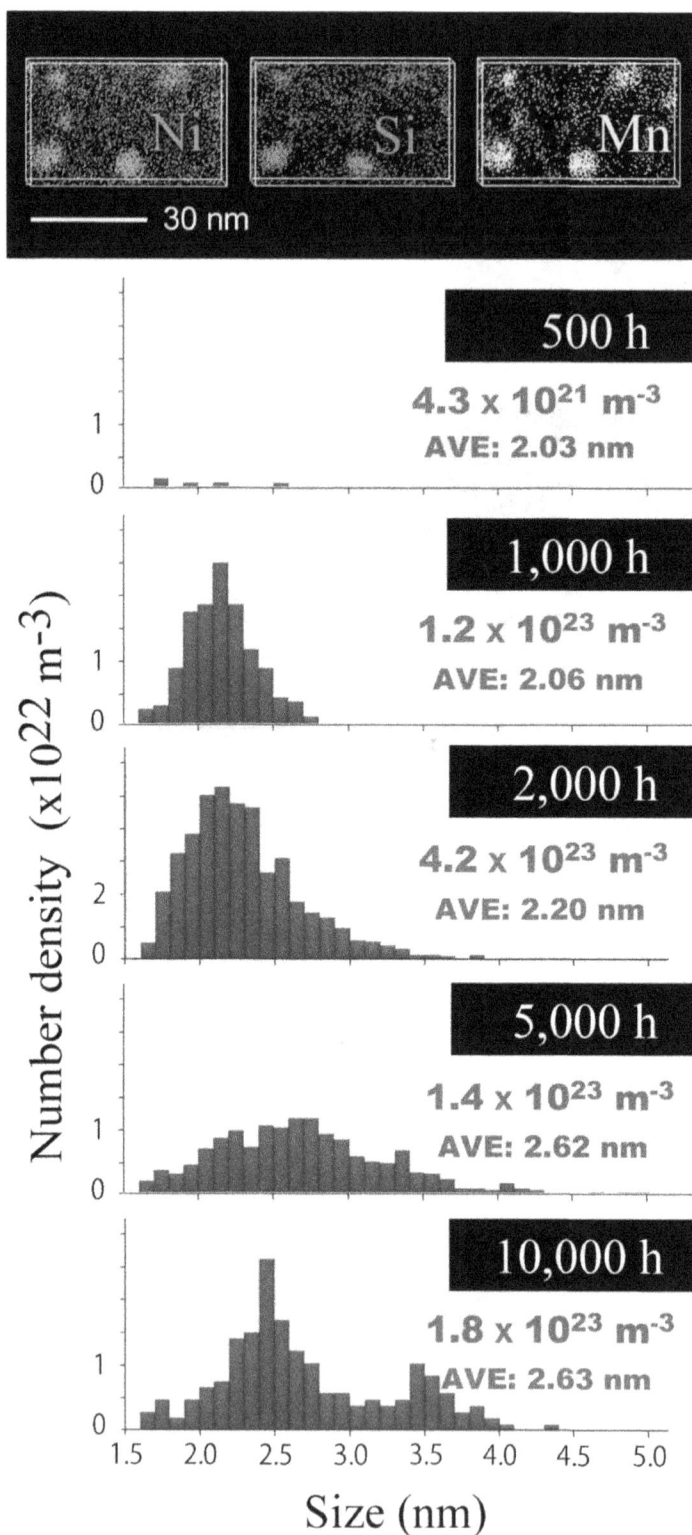

Figure 7
APT results on the steel [3]: size and number density of Ni–Si–Mn clusters.

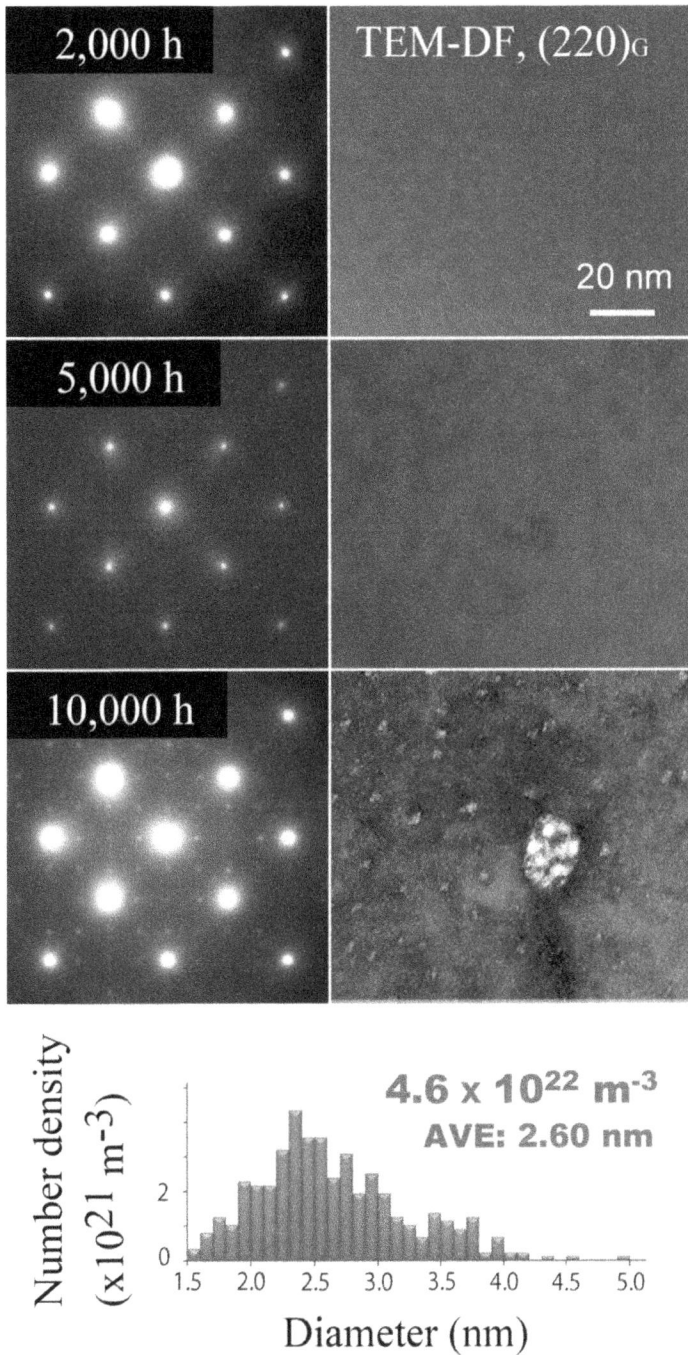

Figure 8
TEM results on the steel [3]. The G-phase precipitates were detected by diffraction pattern and DF image, only in the sample annealed up to 10,000 h. Their number density was only ~26% of the Ni–Si–Mn clusters detected by APT.

The G-phase is currently of particular interest in nuclear materials research, as this compound precipitates also in the steel constituting the main body of reactor pressure vessels (RPVs) at the operation temperature of light water reactors

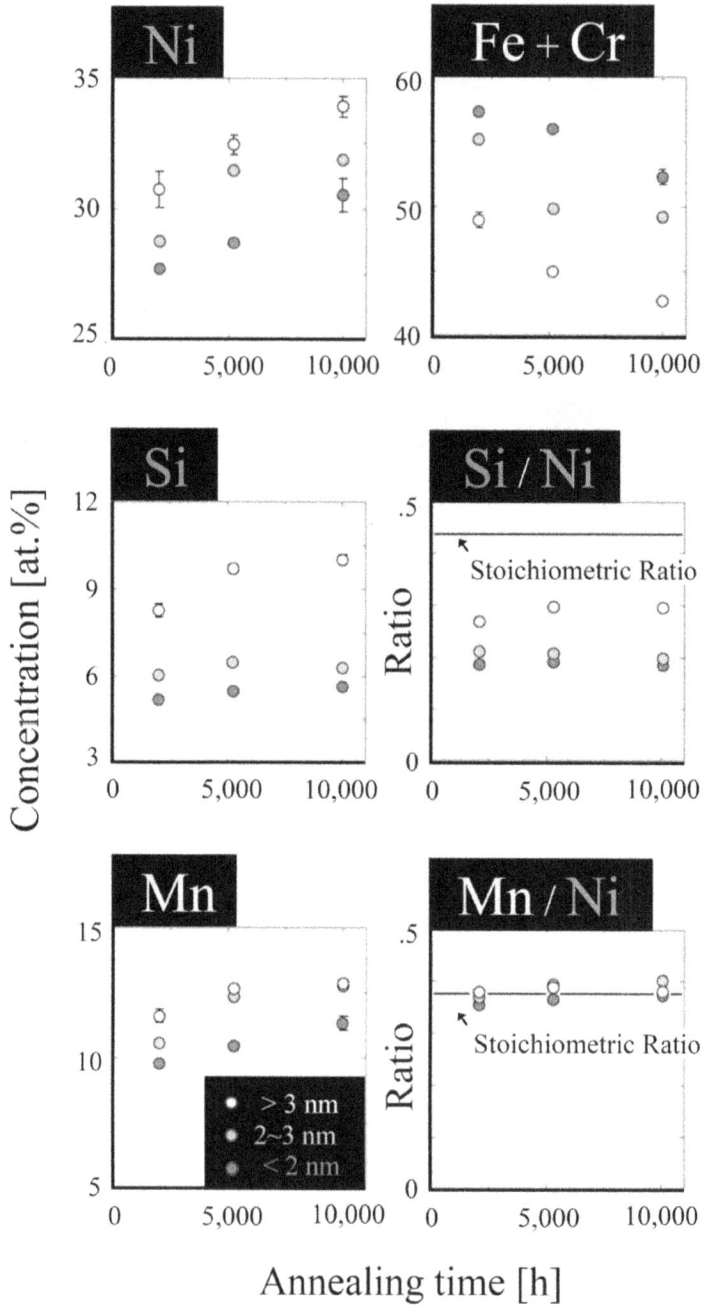

Figure 9
APT results on the steel [3]: composition of Ni–Si–Mn clusters.

(~573 K) under neutron irradiation. It has long been well known that precipitation of impurity Cu causes embrittlement of the RPV steels. In the late 1990s, Odette et al. pointed out that, in the case of RPV steels containing a low amount of Cu such as those manufactured after 1973, precipitation of Cu occurs in the first few years of reactor operation, but near the end of the plants' initial operational license lifetime (typically 40 years), precipitation of Ni, Mn, and occasionally Si becomes dominant over Cu [44]. The Ni–Mn(–Si), precipitates have been called the late-blooming phase [45–47] since their structural and compositional features were unclear at that time. It was very recently that the late-blooming phase is in many cases found to be characterized as the G-phase [48, 49]. The composition of the late-blooming phase detected by APT is not always the same [31, 32]. The composition range of Ni–Si–Mn clusters to become brittle G-phase is a subject to be investigated further.

4. Crystal orientation relationship with the matrix

Crystal orientation relationship between precipitates and matrix is a potential factor controlling the mechanical properties of metals and alloys. Dislocations can glide on specific atomic planes, the choice of which is specific to crystal structure and material. For instance, the slip plane is the {111} plane for fcc metals, the {0001} plane for hcp magnesium, and the {10-10} plane for hcp titanium and hcp zirconium [9]. When the slip plane of precipitates is not parallel to that of the matrix, dislocations are in theory unable to cut through the precipitates. Although the orientation relationship has been extensively studied in the past [50], only a few studies have been reported on the effect of the crystal mismatch on the plasticity [1, 2]. The absence of such studies is partly due to a technical difficulty in determination of crystal orientation of fine precipitate particles as mentioned in the Section 2. Recently, Matsukawa et al. performed a systematic analysis on the parallelism of atomic planes between precipitates and matrix in a Zr–2.5Nb alloy: the precipitates are bcc Nb containing Zr ~10% and the matrix is hcp Zr. Based on the analysis results obtained from 100 precipitate particles (~50 nm in diameter) by means of TKD, they demonstrated that the orientation is practically random. Only 1 out of 100 precipitates had a slip plane parallel to that of the matrix. Their experimental result is inconsistent with a preceding theoretical prediction by Zhang and Kelly [51, 52]. Judging from the mismatch of inter-planar spacings, the most favorable crystal orientation relationship for the Nb-rich bcc precipitates in the hcp Zr matrix is $(\bar{1}011)hcp//(\bar{1}10)_{bcc} \wedge (11\bar{2}3)_{hcp}//(113)_{bcc}$ (**Figure 2**). Matsukawa et al. further demonstrated that the absence of such a specific crystal orientation relationship is attributable to the recrystallization of the matrix. In the Zr–2.5Nb alloy, precipitation occurs in parallel with recrystallization as follows. The Nb atoms are fully dissolved in the matrix at high temperatures with a bcc structure (**Figure 10**). Quenching from this temperature range results in nucleation of bcc Nb nano-precipitates and hcp Zr fine martensites. Ostwald ripening of Nb precipitates occurs during annealing at medium temperatures (773–853 K) together with the recrystallization of the martensite Zr matrix. The initial orientation relationship between the precipitates and the matrix is overwritten by the recrystallization.

In the study of the Zr–2.5Nb alloy, the parallelism of slip planes between precipitates and matrix was analyzed as follows. This analysis is achieved by using the Euler angles obtained from EBSD/TKD measurements, though so far not automated. The analysis procedure is slightly different depending on the analysis software due to the different definition of the Euler angles. In the case of the TSL-OIM software based on Bunge's description [53], the Euler angles (φ_1, Φ, φ_2) are given by three rotations along z_1-x-z_2 axes in accordance with passive rotation (intrinsic rotation), where the axes are rotated instead of the vectors of object, while the object is fixed in space (**Figure 11**). In this case, the rotation matrix (R) relative to the space coordinates is given as follows [54]:

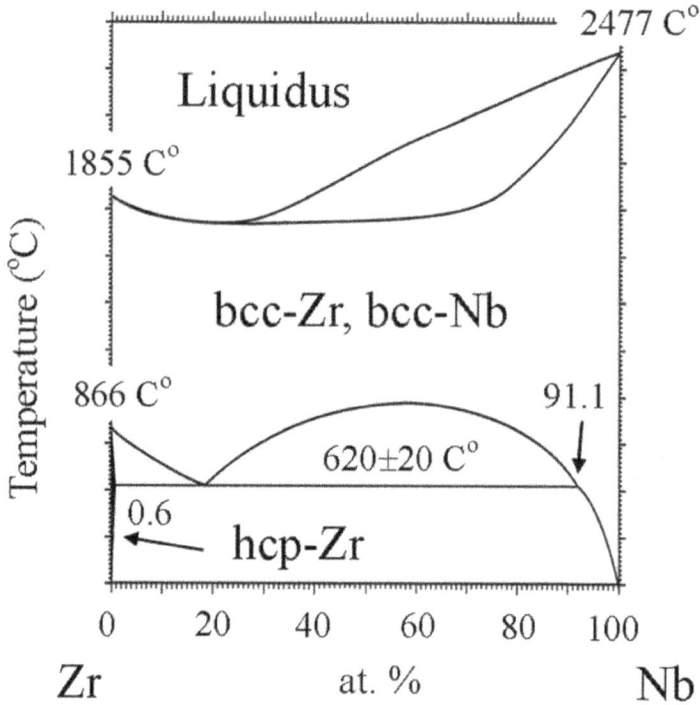

Figure 10
The Zr–Nb binary alloy phase diagram [14].

$$R = R_{Z_2}(\varphi_2)R_x(\Phi)R_{Z_1}(\varphi_1) = \begin{pmatrix} \cos\varphi_2 & \sin\varphi_2 & 0 \\ -\sin\varphi_2 & \cos\varphi_2 & 0 \\ 0 & 0 & 1 \end{pmatrix}\begin{pmatrix} 1 & 0 & 0 \\ 0 & \cos\Phi & \sin\Phi \\ 0 & -\sin\Phi & \cos\Phi \end{pmatrix}\begin{pmatrix} \cos\varphi_1 & \sin\varphi_1 & 0 \\ -\sin\varphi_1 & \cos\varphi_1 & 0 \\ 0 & 0 & 1 \end{pmatrix}$$

$$= \begin{pmatrix} \cos\varphi_1\cos\varphi_2 - \sin\varphi_1\sin\varphi_2\cos\Phi & \sin\varphi_1\cos\varphi_2 + \cos\varphi_1\sin\varphi_2\cos\Phi & \sin\varphi_2\sin\Phi \\ -\cos\varphi_1\sin\varphi_2 - \sin\varphi_1\cos\varphi_2\cos\Phi & -\sin\varphi_1\sin\varphi_2 + \cos\varphi_1\cos\varphi_2\cos\Phi & \cos\varphi_2\sin\Phi \\ \sin\varphi_1\sin\Phi & -\cos\varphi_1\sin\Phi & \cos\Phi \end{pmatrix}$$

$$(1)$$

In the reference crystal, the z- and the x-axes of space coordinates are parallel to the [001] and to the [100] directions of cubic crystals. Here, we consider the orientation relationship between two cubic crystals, A and B, whose rotation matrices relative to the reference crystal are R_A and R_B. The rotation matrix between these two crystals (R_C) is given as follows:

$$\begin{pmatrix} H_1 \\ K_1 \\ L_1 \end{pmatrix} = R_A \begin{pmatrix} 0 \\ 0 \\ 1 \end{pmatrix} \rightarrow R_A{}^{-1}\begin{pmatrix} H_1 \\ K_1 \\ L_1 \end{pmatrix} = \begin{pmatrix} 0 \\ 0 \\ 1 \end{pmatrix} \qquad (2)$$

$$\begin{pmatrix} H_2 \\ K_2 \\ L_2 \end{pmatrix} = R_B \begin{pmatrix} 0 \\ 0 \\ 1 \end{pmatrix} \rightarrow R_B{}^{-1}\begin{pmatrix} H_2 \\ K_2 \\ L_2 \end{pmatrix} = \begin{pmatrix} 0 \\ 0 \\ 1 \end{pmatrix} \qquad (3)$$

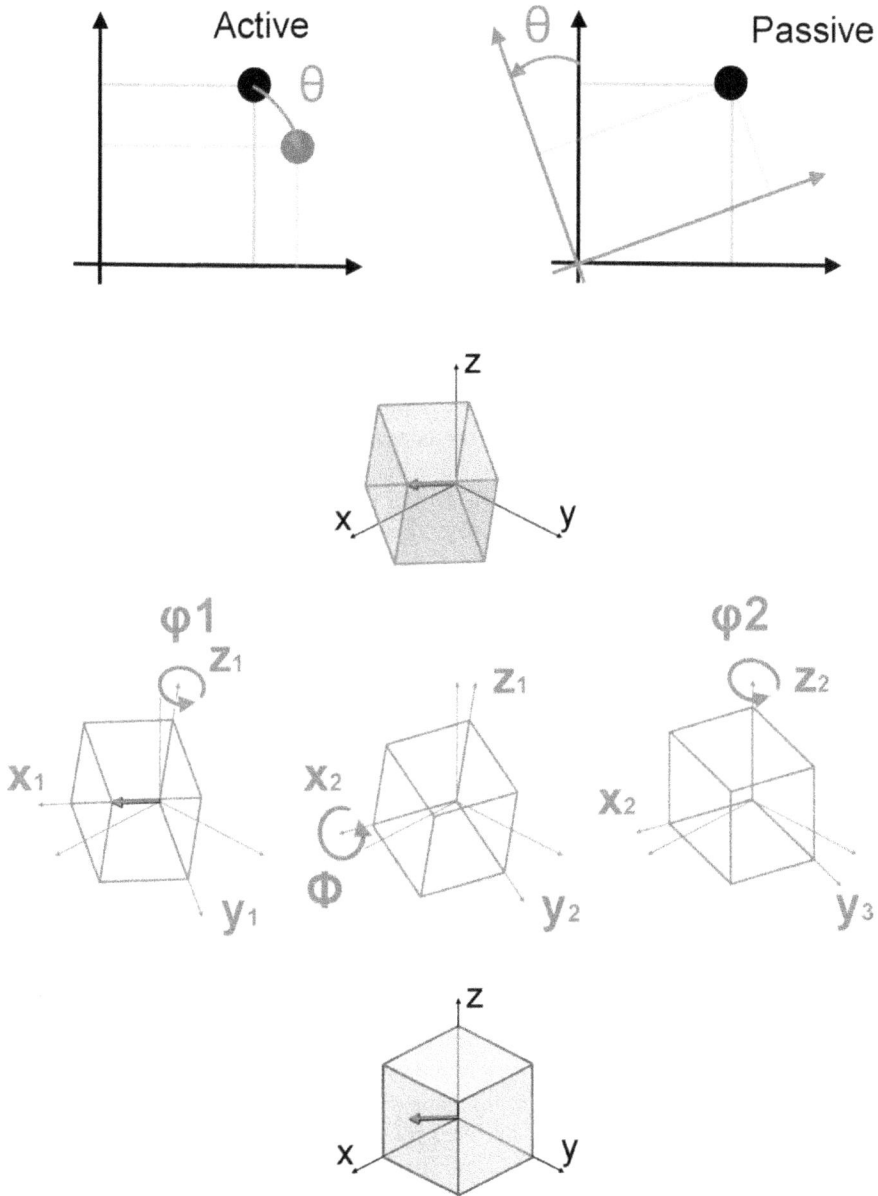

Figure 11
Passive rotations of a cubic crystal with Euler angles (the Bunge Euler angles).

$$\begin{pmatrix} 0 \\ 0 \\ 1 \end{pmatrix} = R_B^{-1} \begin{pmatrix} H_2 \\ K_2 \\ L_2 \end{pmatrix} = R_A^{-1} \begin{pmatrix} H_1 \\ K_1 \\ L_1 \end{pmatrix} \quad \rightarrow \quad \begin{pmatrix} H_2 \\ K_2 \\ L_2 \end{pmatrix} = R_B R_A^{-1} \begin{pmatrix} H_1 \\ K_1 \\ L_1 \end{pmatrix} \quad (4)$$

$$\begin{pmatrix} H_2 \\ K_2 \\ L_2 \end{pmatrix} = R_C \begin{pmatrix} H_1 \\ K_1 \\ L_1 \end{pmatrix} \quad \rightarrow \quad R_C = R_B R_A^{-1} \quad (5)$$

An atomic plane of crystal B, $(H_2' \; K_2' \; L_2')$, parallel to $(H_1' \; K_1' \; L_1')$ of the crystal A is expressed by using the Euler rotation matrix RD, which rotates the sample coordinates in such a way as to match $[H_1' \; K_1' \; L_1']$ to $[001]$ in the space coordinates:

$$\begin{pmatrix} H_1' \\ K_1' \\ L_1' \end{pmatrix} = R_D \begin{pmatrix} 0 \\ 0 \\ 1 \end{pmatrix} \quad \rightarrow \quad \begin{pmatrix} H_2' \\ K_2' \\ L_2' \end{pmatrix} = R_B \, R_A^{-1} \begin{pmatrix} H_1' \\ K_1' \\ L_1' \end{pmatrix} = R_B \, R_A^{-1} R_D \begin{pmatrix} 0 \\ 0 \\ 1 \end{pmatrix}$$

(6)

R_A and R_B are directly determined by EBSD measurements of crystals A and B. The Euler angles of the R_D are determined by using a simulation equipped on the TSL-OIM data collection software. This simulation module is capable of (1) calculating how the index of a crystal (in the ND and the RD directions) changes in accordance with rotations along the three axes and (2) visualizing where the index (of the ND direction) is located on the Kikuchi map (the inverse pole figure). By using these functions, the index $(H_2' \; K_2' \; L_2')$ of precipitate particles can be plotted on one inverse pole figure, though plotting the data points is a time-consuming hand work.

Determination of theoretical accuracy of this analysis method is not straightforward, since errors are introduced by various factors such as (1) the conversion of the Euler angles $(\varphi_1, \Phi, \varphi_2)$ to direction cosines, (2) the conversion of direction cosines denoted in fractional values to the Millar indices (h k l) denoted in integer ratio, and (3) the noise of EBSD data. In order to estimate the practical accuracy of this analysis method, they first analyzed a standard sample in which the atomic-plane parallelism between grains is already known. Their standard sample was a type-316 austenitic stainless steel containing annealing twins (**Figure 12**). The twin boundary of fcc metals is one of the four crystallographically equivalent {111} planes. The Euler angles of these {111} planes for the R_D rotation are, e.g., $(\varphi_1, \Phi, \varphi_2) = (0°, 55°, 45°)$, $(0°, 55°, 135°)$, $(0°, 55°, 225°)$, and $(0°, 55°, 315°)$. They performed this analysis on 50 twin couples and found that the largest offset from the exact {111} was 3.3°. This is the magnitude of practical error of this analysis method.

To date, several orientation relationships have been reported on bcc precipitates in hcp matrix (**Figure 13**). The parallelism of slip planes in those orientation relationships is as follows: (1) the Burgers orientation relationship [13]: $(0001)_{hcp}//(110)_{bcc} \wedge (2\bar{1}\bar{1}0)_{hcp}//(1\bar{1}1)_{bcc}$. The slip plane of hcp Zr matrix, $\{10\bar{1}0\}$, is not exactly parallel to the slip plane of bcc Nb precipitates, {110} or {112}; however, the rotational offset between the $(1\bar{1}0)_{bcc}$ and the $(1\bar{1}00)_{hcp}$ is only 5.3°. (2) The Pitsch-Schrader orientation relationship [55]: $(0001)_{hcp}//(110)_{bcc} \wedge (1\bar{1}00)_{hcp}//(1\bar{1}0)_{bcc}$. The slip planes are exactly parallel to each other. (3) The Potter orientation relationship [56]: $(2\bar{1}\bar{1}0)_{hcp}//(1\bar{1}1)_{bcc}, \wedge (1\bar{1}01)_{hcp}//(011)_{bcc}$. One of the $\{112\}_{bcc}$ is not exactly but nearly parallel to one of the $\{1\bar{1}00\}_{hcp}$. This orientation relationship is close to the Burgers, from which the rotational offset is only $\sim1.5°$. (4) The Rong-Dunlop orientation relationship [57]: $(0001)_{hcp}//(120)_{bcc} \wedge (11\bar{2}0)_{hcp}//(001)_{bcc} \wedge (1\bar{1}00)_{hcp}//(2\bar{1}0)_{bcc}$. The slip plane of hcp and bcc crystals is not parallel to each other. (5) The Zhang and Kelly orientation relationships [51, 52]: $(0001)_{hcp}//(\bar{1}10)_{bcc} \wedge (10\bar{1}0)_{hcp}//(113)_{bcc}$. They proposed several orientation relationships. This one is the most favorable orientation relationship for the bcc Nb precipitates, in terms of the mismatch of the lattice parameter. Their analysis also

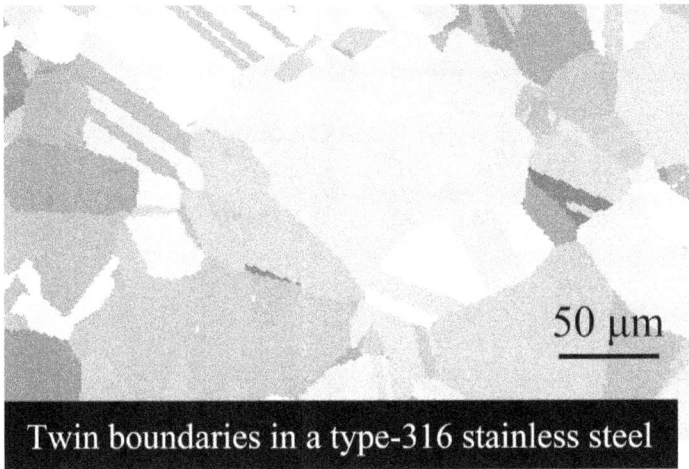

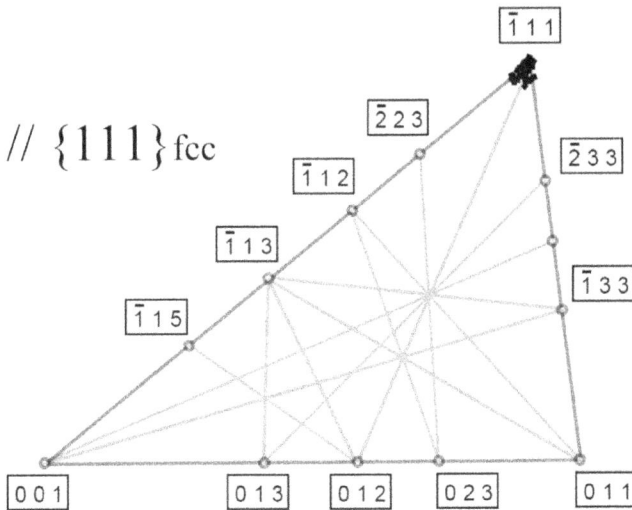

Figure 12
Evaluation of the magnitude of error of the EBSD analysis method on the atomic plane parallelism described in this chapter, using annealing twins in a type-316 stainless steel [2]. The largest offset from the exact {111} was ~3.3°. This is considered as the magnitude of practical error of this analysis method.

suggested that the Pitsch-Schrader and the Rong-Dunlop orientation relationships are favorable over the Burgers orientation relationship for Nb-rich precipitates. This orientation relationship is close to the Burgers, and the slip plane of matrix is not exactly parallel to that of precipitates.

The magnitude of error of the abovementioned analysis of atomic-plane parallelism is greater than the orientation difference between the Potter and the Burgers orientation relationships, 1.5°. It follows that these two orientation relationships are practically indistinguishable from each other in this analysis. On the other hand, the orientation difference between the Pitsch-Schrader and the Burgers orientation relationships is 5.3°; in theory, they are distinguishable. In both the Burgers and the Pitsch-Schrader orientation relationships, the basal plane of the hcp crystal is parallel to a {110} plane of the bcc crystal. In other words, when any one of {110} planes of a precipitate is not parallel to the (0001) plane of the matrix, it follows that the precipitate is in neither one of these two orientation relationships. The criterion for the judgment of whether the Burgers or the Pitsch-Schrader is given by

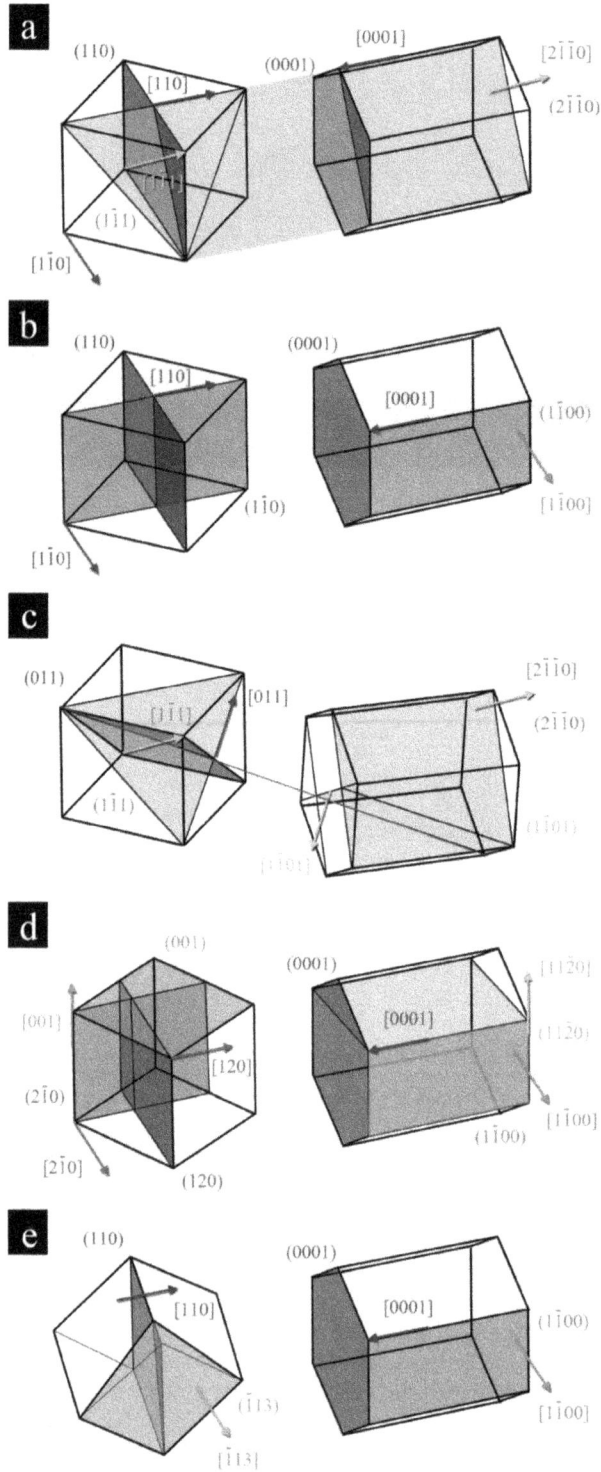

Figure 13
Examples of crystal orientation relationships between bcc and hcp crystals [2]: (a) the Burgers, (b) the Pitsch-Schrader, (c) the Potter, (d) the Rong-Dunlop, and (e) the Zhang-Kelly No. 5.

	Index	Euler angle (φ_1, Φ, φ_2) [°]
Basal	0001	0, 0, 0
Prismatic (type 1)	$01\bar{1}0$	0, 90, 0
	$10\bar{1}0$	0, 90, 60
	$1\bar{1}00$	0, 90, 120
Prismatic (type 2)	$2\bar{1}\bar{1}0$	0, 90, 90
	$1\bar{2}10$	0, 90, 150
	$\bar{1}\bar{1}20$	0, 90, 210

Table 1
The Euler angles to rotate the object coordinates of an hcp crystal in such a way that the plane of interest coincide with the (0001) of the reference hcp crystal, whose [0001] and $[2\bar{1}\bar{1}0]$ are parallel to ND and RD, respectively.

another atomic-plane parallelism, which is whether $\{11\bar{2}0\}_{hcp}//\{111\}_{bcc}$ or $\{1\bar{1}00\}_{hcp}//\{110\}_{bcc}$. As for the parallelism of slip planes, hcp crystals have three crystallographically equivalent $\{10\bar{1}0\}$ planes, whereas bcc crystals have 12 equivalent $\{110\}$ planes and another 12 equivalent $\{112\}$ planes.

In the study of the Zr–2.5Nb alloy, Matsukawa et al. fixed the plane of hcp matrix and plotted its corresponding atomic planes of bcc precipitates on an inverse pole figure (**Figure 4**). In the TSL-OIM software, the Euler angles of hcp crystals are given in the orthogonal coordinate system. In the reference crystal, the z- and the x-axes of space coordinates are parallel to the $[0001]_{hcp}$ and the $[2\bar{1}\bar{1}0]_{hcp}$, respectively.The Euler angles for the RD rotation of the $\{11\bar{2}0\}$ and the $\{10\bar{1}0\}$ planes are shown in **Table 1**.

5. Conclusion

Recent progresses in our understanding of the crystallography of precipitates in metals and alloys have been briefly reviewed. The major highlights are the following three: (1) crystal structure of precipitates changes during nucleation. This concept in itself has been known since the 1930s. Recent new findings concern the critical conditions for the structural change in terms of fluctuations in size and composition, discovered by mean of combining transmission electron microscopy crystallographic analysis with atom probe tomography compositional analysis. It appears that the structural change occurs at a critical size with a critical composition. There is a long incubation period (in some cases a year long) before the structural change after the growth to be the critical size. During the incubation period, enrichment of solute elements occurs inside the precipitates without further size growth. It still remains unclear if these features are universal for any types of precipitates. This research field is expected to advance drastically in the years ahead. (2) In the past years, it has also become technically possible to examine the crystal orientation relationship of fine precipitate particles such as ~50 nm in diameter with the matrix, on numbers of samples numerically sufficient for statistical arguments. Transmission Kikuchi diffraction, which is an advanced technique of electron backscatter diffraction equipped with a scanning electron microscope, revealed that the crystal orientation of precipitates can be random even when they are in theory favorable to have a specific orientation relationship with the matrix from the viewpoint of lattice mismatch. It appears that such a situation is realized when the matrix exhibits recrystallization after precipitation. (3) Crystal orientation relationship between precipitates and matrix was found to be a factor controlling the magnitude of precipitation hardening. This is a new concept beyond the scope of the traditional theory of precipitation hardening, which assumes that the hardening is controlled solely by the shear modulus, whether or not the precipitates are harder than the matrix. In cases where the slip plane of precipitates is not parallel to the slip plane of the matrix, dislocations gliding in the matrix are unable to cut through them, resulting in strong obstacles regardless of the shear modulus. Further information on this issue is provided in the next chapter.

Acknowledgements

The author was supported by the MEXT Grant-in-Aid for Young Scientists (A) (22686058), by the Japan Society for the Promotion of Science (JSPS) KAKENHI (#16K06767), and by the Iron and Steel Institute of Japan (ISIJ) in the 23rd and the 26th Research Promotion Grants. This review article is based on the author's previous researches partly sponsored by the Ministry of Education, Culture, Sports, Science and Technology (MEXT) of Japan, under the Strategic Promotion Program for Basic Nuclear Researches entitled "Study on hydrogenation and radiation effects in advanced nuclear fuel cladding materials" and "Study of degradation mechanism of stainless steel weld-overlay cladding of nuclear reactor pressure vessels" and a program entitled "R&D of nuclear fuel cladding materials and their environmental degradations for the development of safety standards" entrusted to Tohoku University by the MEXT. Those researches were also supported in part by the Collaborative Research Programs of "the Oarai Center" and "the Cooperative Research & Development Center for Advanced Materials" of the Institute for Materials Research, Tohoku University, and of the Research Institute for Applied Mechanics, Kyushu University; by Advanced Characterization Nanotechnology Platform, Nanotechnology Platform Program of the MEXT, Japan, at the Research Center for Ultra-High Voltage Electron Microscopy in Osaka University and at the Ultramicroscopy Research Center in Kyushu University; and by the Joint Usage/ Research Program on Zero-Emission Energy Research, Institute of Advanced Energy, Kyoto University (ZE27C-07, ZE28C-09, ZE29C-11, and ZE30C-01).

Conflict of interest

The author declares no conflicts of interest directly relevant to the content of this chapter.

Author details

Yoshitaka Matsukawa[1,2]

1 Division of Materials Science, Faculty of Advanced Science and Technology, Kumamoto University, Kumamoto, Japan

2 Institute for Materials Research, Tohoku University, Sendai, Japan

*Address all correspondence to: ym2@msre.kumamoto-u.ac.jp

IntechOpen

References

[1] Matsukawa Y, Yang HL, Saito K, Murakami Y, Maruyama T, Iwai T, et al. The effect of crystallographic mismatch on the obstacle strength of second phase precipitate particles in dispersion strengthening: Bcc Nb particles and nanometric Nb clusters embedded in hcp Zr. Acta Materialia. 2016;**102**: 323-332

[2] Matsukawa Y, Okuma I, Muta H, Shinohara Y, Suzue R, Yang HL, et al. Crystallographic analysis on atomic-plane parallelisms between bcc precipitates and hcp matrix in recrystallized Zr–2.5Nb alloys. Acta Materialia. 2017;**126**:86-101

[3] Matsukawa Y, Takeuchi T, Kakubo Y, Suzudo T, Watanabe H, Abe H, et al. The two-step nucleation of G-phase in ferrite. Acta Materialia. 2016;**116**: 104-113

[4] Gibbs JW. On the equilibrium of heterogeneous substances. Transactions Connecticut Academy of Arts and Sciences. **3**(1876):108-248

[5] Orowan E, Zur KI. Zeitschrift für Physik. 1934;**89**:605-613

[6] Polanyi M. Über eine Art Gitterstörung, die einen Kristall plastisch machen könnte. Zeitschrift für Physik. 1934;**89**:660-664

[7] Taylor GI. The mechanism of plastic deformation of crystals. Part I.—Theoretical. Proceedings of the Royal Society of London. Series A. 1934;**145**: 362-387

[8] Hirth JP. A brief history of dislocation theory. Metallurgical Transactions A. 1985;**16A**:2085-2090

[9] Bacon DJ, Osetsky YN, Rodney D. Dislocation–obstacle interactions at the atomic level. In: Hirth JP, Kubin L, editors. Dislocations in Solids.

Amsterdam: Elsevier; 2009. pp. 75, 85-136, 228-237

[10] Orowan E. Discussion. In: Symposium on Internal Stresses in Metals and Alloys, Institute of Metals Monograph and Report Series. Vol. 5. London: Institute of Metals; 1948. pp. 451-453

[11] Ardell AJ. Precipitation hardening. Metallurgical Transactions A. 1985;**16A**: 2131-2165

[12] Mehrer H. Diffusion in Solids. Springer Series Solid State Science. Vol. 155. Berlin; 2007. pp. 297-338

[13] Burgers WG. On the process of transition of the cubic-body-centered modification into the hexagonal-close-packed modification of zirconium. Physica. 1934;**1**:26-586

[14] Okamoto H. Nb–Zr. Journal of Phase Equilibria. 1992;**13**:577-577

[15] Rogers BA, Atkins DF. Zirconium-clumbium diagram. Transactions of AIME. 1944;**203**:1034-1041

[16] Rietveld HM. The rietveld method —A historical perspective. Australian Physics. 1988;**41**:113-116

[17] Yang HL, Shen JJ, Kano S, Matsukawa Y, Li YF, Satoh Y, et al. Effects of Mo addition on precipitation in Zr–1.2Nb alloys. Materials Letters. 2015;**158**:88-91

[18] Tanigawa H, Sakasegawa H, Klueh RL. Irradiation effects on precipitation in reduced-activation ferritic/martensitic steels. Metallurgical Transactions. 2005;**46**:469-474

[19] Buseck PR, Iijima S. High resolution electron microscopy of silicates. American Mineralogist. 1974;**59**:1-21

[20] Rauch EF, Portillo J, Nicolopoulos S, Bultreys D, Rouvimov S, Moeck P. Automated nanocrystal orientation and phase mapping in the transmission electron microscope on the basis of precession electron diffraction. Zeitschrift für Kristallographie. 2010;**225**:103-109

[21] Viladot D, Véron M, Gemmi M, Peiró F, Portillo J, Estradé S, et al. Orientation and phase mapping in the transmission electron microscope using precession-assisted diffraction spot recognition: State-of-the-art results. Journal of Microscopy. 2013;**252**:23-34

[22] Williams DB, Carter CB. Transmission Electron Microscopy. New York: Springer; 1996. pp. 291-299

[23] Zaefferer S. A critical review of orientation microscopy in SEM and TEM. Crystal Research and Technology. 2011;**46**:607-628

[24] Trimby PW. Orientation mapping of nanostructured materials using transmission Kikuchi diffraction in the scanning electron microscope. Ultramicroscopy. 2012;**120**:16-24

[25] Rice KP, Keller RR, Stoykovich MP. Specimen-thickness effects on transmission Kikuchi patterns in the scanning electron microscope. Journal of Microscopy. 2014;**254**:129-136

[26] Hirsch PB, Howie A, Whelan MJ. Electron microscopy of thin metals. 2nd ed. Malabar: R.E. Krieger Publishing Company; 1977. pp. 119-121

[27] Gault B, Moody MP, Cairney JM, Ringer SP. Atom probe tomography. In: Springer Series in Materials Science. Vol. 160. New York: Springer; 2011. pp. 185-190

[28] Miller MK, Cerezo A, Hetherington MG, Smith GDW. Atom probe field ion microscopy. In: Monographs on the Physics and Chemistry of Materials. Vol. 52. Oxford: Oxford Science Publications; 1999. pp. 191-199

[29] Blavette D, Vurpillot F, Pareige P, Menand A. A model accounting for spatial overlaps in 3Datom-probe microscopy. Ultramicroscopy. 2001;**89**: 145-153

[30] Vurpillot F, Bostel A, Blavette D. Trajectory overlaps and local magnification in three-dimensional atom probe. Applied Physics Letters. 2000;**76**:3127-3129

[31] Kuramoto A, Toyama T, Takeuchi T, Nagai a Y, Hasegawa M, Yoshiie T, et al. Post-irradiation annealing behavior of microstructure and hardening of a reactor pressure vessel steel studied by positron annihilation and atom probe tomography. Journal of Nuclear Materials. 2012;**425**:65-70

[32] Kuramoto A, Toyama T, Nagai Y, Inoue K, Nozawa Y, Hasegawa M, et al. Microstructural changes in a Russian-type reactor weld material after neutron irradiation, post-irradiation annealing and re-irradiation studied by atom probe tomography and positron annihilation spectroscopy. Acta Materialia. 2013;**61**:5236-5246

[33] Erdemir D, Lee AY, Myerson AS. Nucleation of crystals from solution: Classical and two step models. Accounts of Chemical Research. 2009;**42**:621-629

[34] Peng Y, Wang F, Wang Z, Alsayed AM, Zhang Z, Yodh AG, et al. Two-step nucleation mechanism in solid–solid phase transitions. Nature Materials. 2014;**14**:101-108

[35] Borelius G. Zur theorie der umwandlungen von metallischen mischphasen. Annals of Physics. 1937; **28**:507-519

[36] Hobstetter JN. Stable transformation nuclei in solids. Transactions of the AIME. 1949;**180**:12130

[37] Ostwald W. The formation and changes of solids (Translated from German). Zeitschrift für Physikalische Chemie. 1897;**22**:289-330

[38] Chung SY, Kim YM, Kim JG, Kim YJ. Multiphase transformation and Ostwald's rule of stages during crystallization of a metal phosphate. Nature Physics. 2009;**5**:68-73

[39] Biswas A, Siegel DJ, Wolverton C, Seidman DN. Precipitates in Al–Cu alloys revisited: Atom-probe tomographic experiments and first-principles calculations of compositional evolution and interfacial segregation. Acta Materialia. 2011;**59**:6187-6204

[40] Marceau RKH, Sha G, Ferragut R, Dupasquier A, Ringer SP. Solute clustering in Al–Cu–Mg alloys during the early stages of elevated temperature ageing. Acta Materialia. 2010;**58**: 4923-4939

[41] Othen PJ, Jenkins ML, Smith DW. High-resolution electron microscopy studies of the structure of Cu precipitates in α-Fe. Philosophical Magazine: A. 1994;**70**:1-24

[42] Mateo A, Llanes L, Anglada M, Redjaimia A, Metauer G. Characterization of the intermetallic G-phase in an AISI 329 duplex stainless steel. Journal of Materials Science. 1997; **32**:4533-4540

[43] Miller MK, Bentley J. APFIM and AEM investigation of CF8 and CF8M primary coolant pipe steels. Journal of Materials Science and Technology. 1990;**6**:285-292

[44] Zinkle SJ, Was GS. Materials challenges in nuclear energy. Acta Materialia. 2013;**61**:735-758

[45] Odette GR, Wirth BD. A computational microscopy study of nanostructural evolution in irradiated pressure vessel steels. Journal of Nuclear Materials. 1997;**251**:157-171

[46] Soneda N. In: Soneda N, editor. Irradiation Embrittlement of Reactor Pressure VesselsEmbrittlement correlation methods to identify trends in embrittlement in reactor pressure vessels (RPVs). Cambridge: Woodhead Publ Ltd; 2015. pp. 333-377

[47] Ngayam-Happy R, Becquart CS, Domain C, Malerba L. Formation and evolution of MnNi clusters in neutron irradiated dilute Fe alloys modelled by a first principle-based AKMC method. Journal of Nuclear Materials. 2012;**426**: 198-207

[48] Wells PB, Yamamoto T, Miller B, Milot T, Cole J, Wu Y, et al. Evolution of manganese–nickel–silicon-dominated phases in highly irradiated reactor pressure vessel steels. Acta Materialia. 2014;**80**:205-219

[49] Sprouster DJ, Sinsheimer J, Dooryhee E, Ghose SK, Wells P, Stan T, et al. Structural characterization of nanoscale intermetallic precipitates in highly neutron irradiated reactor pressure vessel steels. Scripta Materialia. 2016;**113**:18-22

[50] Dahmen U. Orientation relationships in precipitation systems. Acta Metallurgica. 1982;**30**:63-73

[51] Zhang M-X, Kelly PM. Crystallographic features of phase transformations in solids. Progress in Materials Science. 2009;**54**: 1101-1170

[52] Zhang M-X, Kelly PM. Edge-to-edge matching and its applications: Part I. Application to the simple HCP/BCC system. Acta Materialia. 2005;**53**: 1073-1084

[53] Bunge H-J. Texture Analysis in Materials Science: Mathematical

Methods. English ed. London:
Butterworth; 1982. pp. 4-22

[54] Engler O, Landle V. Introduction to
Texture Analysis. 2nd ed. New York:
CRC Press; 2010. pp. 33-36

[55] Pitsch W, Schrader A. Die
ausscheidungsform des ε-karbids im
ferrit und im martensit beim anlassen.
Archiv fur das Eisenhuttenwesen. 1958;
29:715-721

[56] Potter DL. The structure,
morphology and orientation
relationship of V_3N in α-vanadium.
Journal of the Less Common Metals.
1973;**31**:299-309

[57] Rong W, Dunlop GL. The
crystallography of secondary carbide
precipitation in high speed steel. Acta
Metallurgica. 1984;**32**:1591-1599

Chapter 3

Crystallography of Precipitates in Metals and Alloys: (2) Impact of Crystallography on Precipitation Hardening

Yoshitaka Matsukawa

Abstract

Following the previous chapter, this chapter describes crystallography of second-phase precipitate particles in metals and alloys; the focus of this chapter is placed on the effect of crystallography of precipitates on precipitation hardening. Unlike nonmetallic composite materials whose strength is determined by the volume fraction ratio of constituent phases, the strength of metals and alloys can be several times greater by introducing a minor amount of precipitate particles such as 2%. The magnitude of strengthening (hardening) due to precipitates is, in traditional understanding, controlled by the shear modulus, whether or not the precipitates are harder than the matrix. The most recent major update in this research field is a discovery that crystallography of precipitates is another factor controlling the magnitude of strengthening. In the case where the slip plane of dislocations in precipitates is not parallel to that in the matrix, dislocations gliding in the matrix are unable to cut through the precipitates, resulting in intense hardening regardless of the shear modulus. This chapter also reviews the classical theory of precipitation hardening established in the 1950s–1960s, in order to sort out open questions to be resolved.

Keywords: precipitates, nucleation, crystal structure, strength, dislocations

1. Introduction

This chapter is a supplement to the previous chapter on crystallography of precipitate particles in metals and alloys, for the purpose of describing how the crystallography of precipitates practically affects the physical properties of entire the material. The crystallography of precipitates is of interest not only for fundamental materials science but also for engineering, in particular, structural materials engineering. The strength of metals and alloys is highly affected by a minor amount of precipitates such as a few percent. In the case of nonmetallic composite materials, their strength is determined by the volume fraction ratio of constituent phases. In other words, the strength of nonmetallic composites is expected not to exceed that of constituent phases. On the other hand, metals and alloys containing second-phase precipitate particles, say, 2% in volume fraction, can exhibit a strength several times greater than the matrix phase (**Figures 1–3**). Such intense hardening is

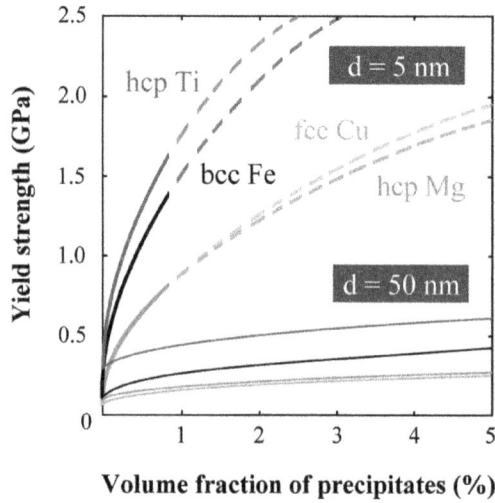

Figure 1.
A model calculation of precipitation hardening in the hcp Ti, the hcp Mg, the fcc Cu, and the bcc Fe, as a function of the volume fraction of precipitates for the cases of precipitate diameter of 5 and 50 nm. The obstacle strength is set to α = 0.8.

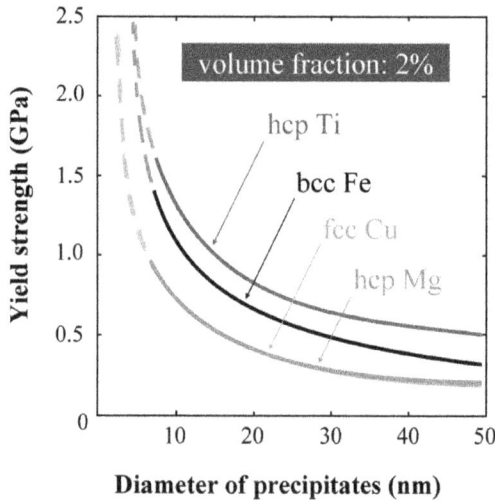

Figure 2.
A model calculation of precipitation hardening in the hcp Ti, the hcp Mg, the fcc Cu, and the bcc Fe, as a function of the diameter of precipitates for the case of precipitate volume fraction of 2%. The obstacle strength is set to α = 0.8.

achieved when precipitate particles are strong obstacles against the motion of dislocations gliding on a slip plane in the matrix. They are strong obstacles in the case where dislocations are unable to cut through them (**Figure 4**). In the classical theory of precipitation hardening (a.k.a. dispersion strengthening) established in the 1950s–1960s, the obstacle strength is assumed to be determined by the shear modulus [1, 2]; those which are harder than the matrix are strong obstacles. In general, this condition is fulfilled by a combination of metallic matrix and nonmetallic compound precipitates such as oxides and carbides whose strength is typically a few GPa, which is ~10 times greater than the yield strength of metals. Recent experimental studies demonstrated that crystallography of precipitate particles is

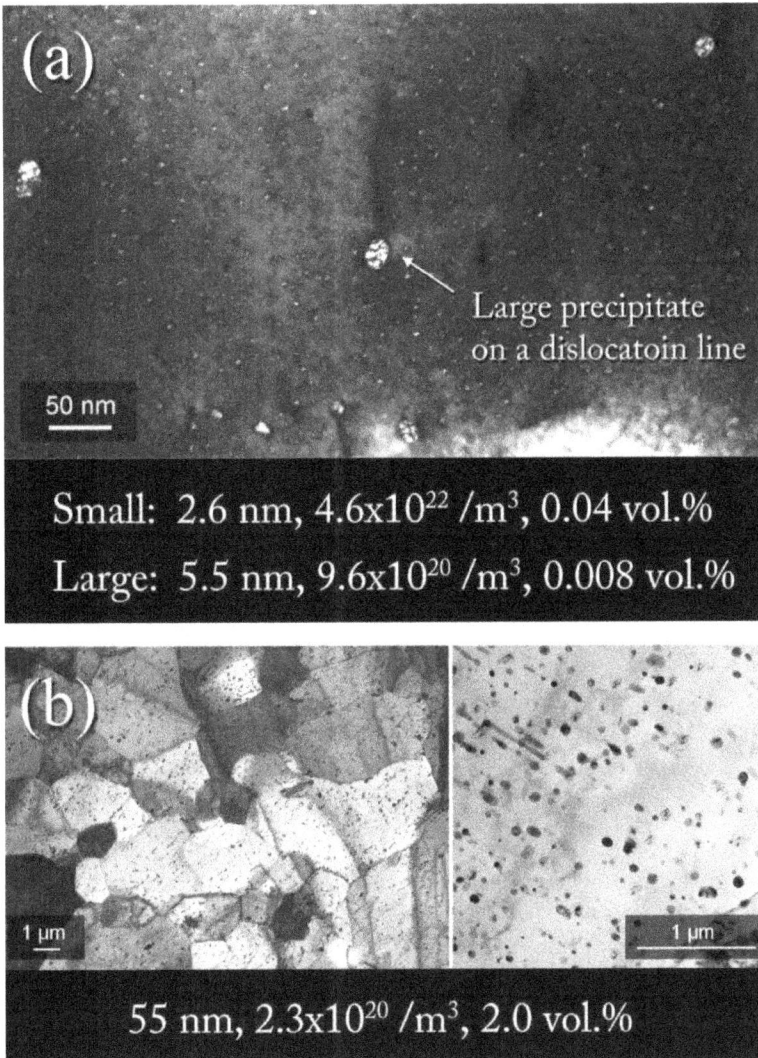

Figure 3.
Example of precipitate particles in alloys: (a) G-phase precipitates in a duplex stainless steel and (b) bcc Nb precipitates in a Zr—2.5Nb alloy [4].

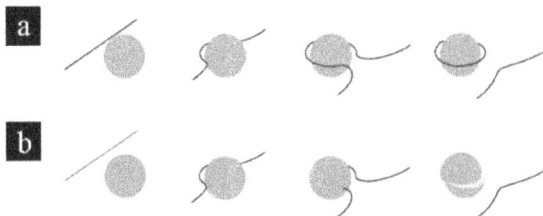

Figure 4.
Interaction between a gliding dislocation and a precipitate particle: (a) the Orowan mechanism for strong obstacles and (b) the cutting mechanism for weak obstacles [1]. The factor controlling the obstacle strength has been assumed to be the shear modulus, i.e., precipitates harder than matrix are strong obstacles. This concept has recently been updated; crystallography of precipitates is another factor controlling their obstacle strength.

another factor dominating their obstacle strength [3, 4]. When the slip plane in precipitates is not parallel to that in the matrix, dislocations gliding in the matrix are unable to cut through the precipitates regardless of the shear modulus. In this case, metallic precipitates are strong obstacles as well as nonmetallic compound precipitates. It follows that metals and alloys can be several times stronger than their constituent phases. In this way, crystallography enables us to create strong materials from a combination of weak materials: one plus one becomes more than two.

2. Magnitude of hardening as a function of precipitate size and number density

The abovementioned statement about the magnitude of precipitation hardening of the case of 2% volume fraction is derived from the following numerical calculation. Based on a geometry consideration of dislocation-precipitate interaction, the increase of material's yield strength, σ_y, which is a critical stress level where the deformation mode changes from elastic to plastic deformation, is given as follows [3, 5–7]:

$$\Delta\sigma_y = \alpha M \mu b (N_v d)^{1/2} \tag{1}$$

where α is the obstacle strength of the precipitates, M is the Taylor factor, μ is the shear modulus of the matrix, b is the magnitude of the Burgers vector of dislocations in the matrix, and N_v and d are the number of precipitate particles per unit volume (i.e., the number density) and their mean diameter. The Taylor factor (M) is a material-specific constant primarily dependent on the crystal structure, slip systems, and texture. An M value of 3.1 is commonly applied to non-textured polycrystalline metals having an fcc structure or a bcc structure [5–7]. For metals having an hcp structure, M values of 6.5 and 5.0 are commonly applied to Mg [8, 9] and Ti [10, 11], respectively. Their difference is related to the number of active slip system, which is dependent on the c/a ratio. The c/a ratio is 1.633 for ideal close-packed structure, 1.623 for Mg, and 1.587 for Ti. In the hcp Mg only the basal plane is available for dislocation slip, whereas in the hcp Ti, the primary slip plane is the prism plane, and the basal plane is also available as a secondary slip plane. In the traditional concept of the Orowan hardening, the obstacle strength (α) is dependent on the shear modulus. The obstacle strength of strong obstacles is, in theory, $\alpha = 1$. However, when particles are dispersed in random distribution, the obstacle strength α is no longer 1, but instead a factor of 0.80–0.85 is introduced [12–16].

In the past several years, Eq. (1) has been frequently cited especially in the research community of nuclear materials, which exhibit hardening and embrittlement due to precipitation induced by high-energy neutron irradiation. According to previous publications [5–7], the source of this equation goes back to a paper published in 1958 [17]. This information appears to be incorrect; Eq. (1) is not mentioned in it. For future reference, here we provide the detailed derivation method of Eq. (1) as follows.

When a gliding dislocation is pinned by a precipitate particle (**Figure 5**), the force acting on the dislocation (F) is given as a function of bowing angle (θ) and the line tension (T):

$$F = 2T\cos(\theta/2) \tag{2}$$

The force acting on a dislocation line is alternatively given as a function of a shear stress (τ), the Burgers vector (b), and the length of dislocation line, which is in this case the distance between particles (L):

$$F = \tau bL \rightarrow \tau = \frac{2T}{bL} \cos{(\theta/2)} \tag{3}$$

By adopting the simplest approximation for the line tension of dislocations, i.e., $T = \mu b^2/2$, we obtain:

$$\tau = \frac{\mu b}{L} \cos{(\theta/2)}(\rightarrow= \alpha\mu b/L) \tag{4}$$

The obstacle strength (α) corresponds to the bowing angle (θ). In the case where the precipitate particle is an impenetrable obstacle, the dislocation eventually bypasses the obstacle with forming a dislocation loop around it (**Figure 4**). The formation of dislocation loop occurs at $\theta = 0°$, where the segments of dislocation line on both sides of the obstacle become parallel from each other. Those segments have the same Burgers vector and opposite line senses, thereby they are attracted each other and eventually merged into one. At $\theta = 0°$, F becomes a maximum value (=2 T). In other words, the maximum value of α is 1. Degree of precipitation hardening in macroscopic length scale is denoted by the tensile stress rather than the shear stress; they converted each other using the Taylor factor ($\tau = \sigma/M$):

$$\Delta\sigma_y = \frac{M\mu b}{L} \cos{(\theta/2)}(\rightarrow= \alpha M\mu b/L) \tag{5}$$

Eq. (5) is converted into Eq. (1), based on the simplest approximation for the spatial distribution of precipitates, i.e., a square lattice arrangement:

$$L = 1/N_s^{1/2} = 1/(2rN_v)^{1/2} = 1/(N_v d)^{1/2} \tag{6}$$

where N_s is the number of precipitate particles per unit area on a plane (i.e., the planar number density) and r is the mean radius of precipitates. The relation of $N_s = 2rN_v$ is derived from the Delesse's principle [18], where the volume fraction of precipitate particles in 3D space (V_V) is equivalent to their area fraction in 2D space (A_A):

$$V_V = A_A$$

$$N_v \times \left(\frac{4\pi r^3}{3}\right) = N_s \times \pi(r')^2$$

$$N_v \times \left(\frac{4\pi r^3}{3}\right) = N_s \times \pi\left(\frac{r}{1.22}\right)^2$$

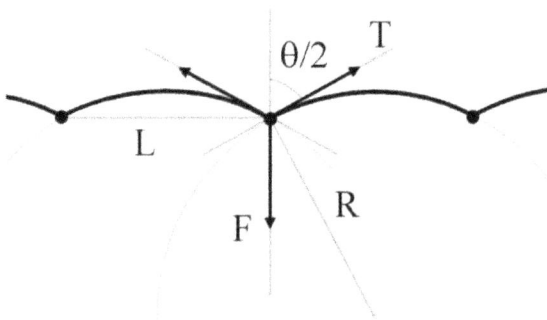

Figure 5.
Dislocation on a slip plane where each obstacle exerts localized glide resistance force (F) balanced in equilibrium by line tension forces (T).

$$N_s \approx 2rN_v \qquad (7)$$

The average radius of particles measured on 2D space (r') is a function of their true radius (r): r' = r/1.22. This relationship is derived as follows [19]. The radius (r') varies with the position of sectioning plane relative to the center of sphere:

$$r' = \sqrt{r^2 - h^2} \qquad (8)$$

where h is the distance from the center to the sectioning plane. The probability that the test plane intersects the sphere at a distance between h and h + Δh from its center is dh/r. The average of the area of section $\overline{\pi(r')^2}$ is obtained by applying the definition of mean value [19] as follows:

$$\overline{\pi(r')^2} = \int_0^r \pi(r^2 - h^2)\frac{dh}{r} = \frac{2}{3}\pi r^2$$

$$(r')^2 = \frac{2}{3}r^2$$

$$r \approx 1.2247r' \qquad (9)$$

The obstacle strength α is 1 for particles dispersed in an ideal square lattice arrangement but ∼0.8 for those dispersed randomly. This is an empirical rule obtained from 2D simulations performed by Foreman and Makin (**Figure 6**). The motion of dislocations is not uniform, but they propagate preferentially through "local soft spots" where the local number density of obstacles is smaller than the others (**Figure 7**). As a result of such spatially localized deformation, the average stress level required to sweep a unit area becomes smaller than the case of square arrangement.

An alternative explanation of α ≠ 1 is obtained from a geometry analysis (**Figure 8**) considering the possibility that effective average interparticle distance (Λ) on the slip plane may be different from the actual average interparticle distance (L) in 3D space due to the effect of a dislocation-obstacle interaction on another interaction. This concept is based on an assumption that the lattice friction against dislocation glide is negligibly small, in a steady-state plastic deformation at a

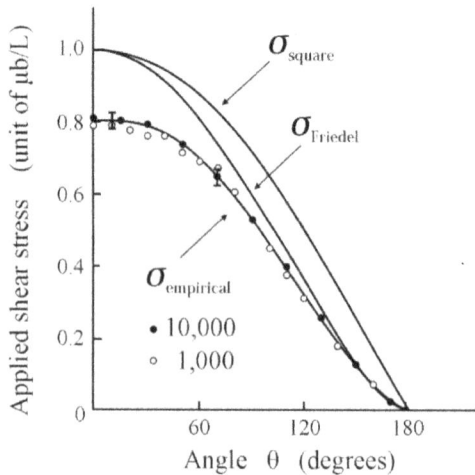

Figure 6.
*Simulation results of Foreman and Makin [12] in comparison with the Orowan model, which assumes a square lattice arrangement of obstacles, and a calibrated model based on the Friedel's assumption shown in **Figure 14**.*

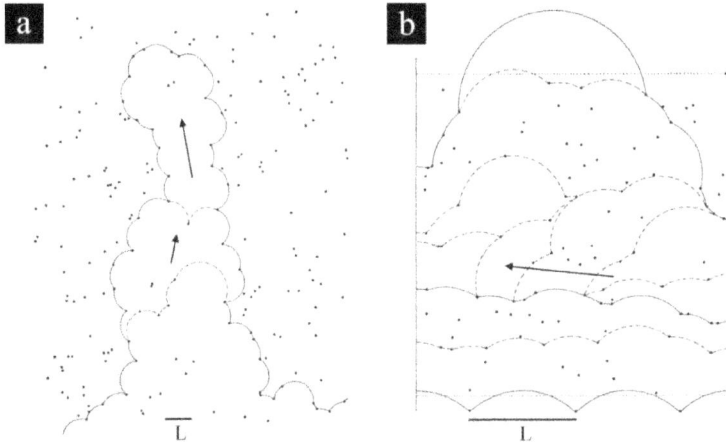

Figure 7.
Simulation results of Foreman and Makin [12]: (a) strong obstacles and (b) weak obstacles. The arrows indicate the propagation direction of dislocations.

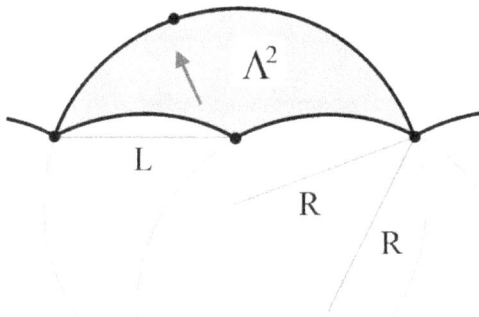

Figure 8.
A geometry consideration of simultaneous interaction of a dislocation with multiple obstacles. When a dislocation segment starts interacting with an obstacle, this segment is already curved due to the aftereffect of previous interaction with another obstacle. Since the dislocation segment sweeps an area between obstacles without any additional stress (when the friction is assumed to be negligibly small), the average flow stress becomes smaller that of the case of single obstacle interaction (and also the square lattice arrangement). Consequently, their effective obstacle strength becomes smaller.

constant strain rate. In this case, after unpinning from an obstacle, the curvature of a dislocation remains unchanged until it encounters another obstacle [20] (**Figure 8**). If the area swept by such motion of dislocations upon unpinning is greater than the area occupied by one obstacle in the average of 3D space (given as a regular square lattice arrangement), it follows that the apparent stress level required to sweep a unit area becomes smaller. The area swept by the dislocation (Λ^2) is regarded as the effective area occupied by a single particle on the slip plane. This area is the area of large segment of a circle of radius (R) minus that of two small ones:

$$\Lambda^2 = \left[R^2 \sin^{-1}\left(\frac{L}{2R}\right) - \frac{L}{4}\sqrt{4R^2 - L^2} \right] - 2\left[R^2 \sin^{-1}\left(\frac{L}{R}\right) - L\sqrt{R^2 - L^2} \right] \quad (10)$$

This formula cannot be solved unless otherwise some kind of approximation is introduced. A common approximation of L < <R results in $\Lambda^2 = 0$. Judging from the geometry shown in **Figure 8**, a rather approximation is L = R.

$$\Lambda^2 = \left(\frac{\pi}{6} + \frac{\sqrt{3}}{2}\right)L^2$$

$$\Lambda \approx 1.18L \tag{11}$$

In this case, a factor of $1/1.18 = 0.85$ is introduced into Eq. (4). In other words, the maximum value of α is 0.85 if this geometry effect is considered. It is noteworthy that this value coincides with the coefficient of the well-known Ashby-Orowan model [16]. The Ashby-Orowan model adopted this value from Kocks' works, which consist of a graphical analysis performed on 550 obstacles in random distribution [13] and a geometry consideration expressed in a more complicated form than this analysis [15].

Eq. (1) is, although obtained from simplified assumptions in terms of the line tension of dislocations and the spatial distribution of precipitates, helpful for intuitive understanding about the effects of precipitate size and number density on hardening. Using this formula, here we evaluate the magnitude of hardening as a function of those factors in fcc Cu, bcc Fe, hcp Ti, and hcp Mg. Under a constant volume fraction of precipitates, smaller size results in greater hardening due to:

$$V_v = N_v \times \left(\frac{4\pi r^3}{3}\right) \tag{12}$$

Smaller size results in higher number density. In many cases, the maximum value of the number density of precipitates introduced by thermal aging is $\sim 1 \times 10^{23}$ m^{-3}, which corresponds to an interparticle distance of ~ 22 nm in the case of the square lattice distribution. With this number density, when the volume fraction is 2%, it follows that the diameter of particles is 7.3 nm (**Table 1**). In the case of bcc Fe, the yield strength is estimated to be ~ 1.4 GPa, ~ 10 times greater than the yield strength without precipitates, ~ 120 MPa. Irradiation with high-energy particles such as neutron often induces (or enhance) precipitation of second phase. The number density of irradiation-induced precipitates can become the order of 10^{24}/m^3, which corresponds to an interparticle distance of 10 nm. For instance, the number density and diameter of neutron irradiation-induced Cu precipitates and Ni—Si—Mn precipitates are both some 10^{24}/m^3 and 0.5–1.5 nm, respectively [21]. Their volume fractions are 0.007% for 0.5 nm and 0.18% for 1.5 nm. Their hardening is estimated to be 1.1 GPa for the former and 1.9 GPa for the latter. For reference, the magnitude of hardening in the bcc Fe, the fcc Cu, the hcp Ti, and the hcp Mg as a function of volume fractions and diameter of precipitates is summarized in **Figures 1** and **2**.

Number density [m^{-3}]	Volume fraction [%]	Diameter [nm]	Interparticle distance [nm]	$\Delta\sigma$ [MPa]
2×10^{23}	2	5.8	17.1	1750
1×10^{23}	2	7.3	21.5	1365
1×10^{22}	2	15.6	46.4	631
1×10^{21}	2	34	100	295
1×10^{20}	2	73	215	137

Table 1.
Estimation of realistically achievable maximum precipitation hardening in bcc iron at a constant volume fraction (2%) with variation of number density and mean diameter of precipitate particles.

3. Effects of crystallography on obstacle strength

After Orowan, extensive studies have been made on the effects of various factors such as dislocation character (edge vs. screw), spatial distribution of precipitates and their size distribution, elastic anisotropy, stacking fault energy, coherency, formation of ledges at the precipitate/matrix interface due to passage of dislocations (a.k.a. chemical strengthening), formation of antiphase boundary at the interface (a.k.a. ordering strengthening), etc. However, the effect of crystallography of precipitates has long been unexplored until very recently, partly due to technical difficulties in experiments. The absence of simulations on this issue is due to the following two reasons. (1) For molecular dynamics (MD) simulations, reproducing the realistic interaction geometry between gliding dislocations and incoherent particles is technically rather difficult, because experimental databases on the atomic structures of the precipitate-matrix interphases are limited. (2) For dislocation dynamics (DD) simulations based on continuum elasticity theory calculations, the effect of crystal mismatch is beyond the capability.

In 2016 it was experimentally demonstrated that soft precipitates can be strong obstacles. That report examined bcc Nb precipitates in hcp Zr matrix, the shear modulus of which are 28 and 33 GPa, respectively. Traditionally, the obstacle strength of such soft precipitates has been scaled by the difference of the shear modulus between precipitates and matrix in accordance with a model proposed by Russell and Brown in 1972 [22]. An implication of the Russell-Brown model is that a greater difference in the shear modulus results in a greater obstacle strength, as described later. Since the Nb precipitates in the Zr—2.5Nb alloy are as soft as the Zr matrix, they are considered weak obstacles. Nevertheless, their experimentally determined obstacle strength was α = 0.8–1 (**Figure 9**), indicating that they are ideal Orowan-type strong obstacles. This analysis result is supported by transmission electron microscopy (TEM) observation (**Figure 10**). The morphology of the Nb precipitates does not change even after severe cold rolling up to 90%; they are certainly non-shearable obstacles. Later, Matsukawa et al. further demonstrated that, by means of transmission Kikuchi diffraction (TKD), crystal orientation of the Nb precipitates is practically random (as described in the previous section of this chapter). Considering that dislocations can glide only on specific atomic planes, the most probable scenario is probably that dislocations cannot cut through the

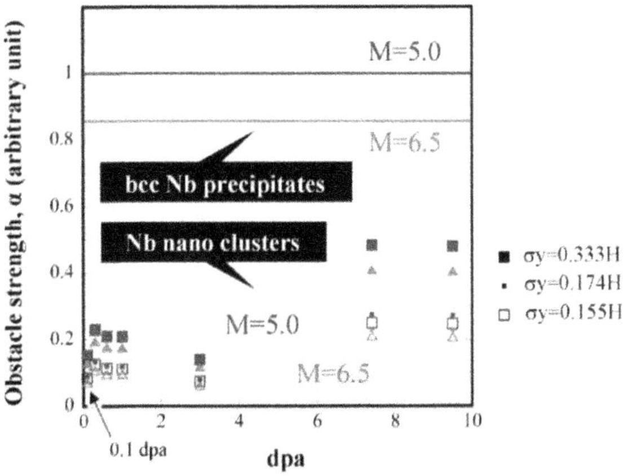

Figure 9.
Experimentally determined obstacle strength of bcc Nb precipitates and irradiation-induced Nb nanoclusters in Zr—Nb alloys [3]. The former was obtained from Zr—Nb alloys containing various amounts of bcc Nb precipitates, whereas the latter from a Zr—2.5Nb alloy subjected to ion irradiation.

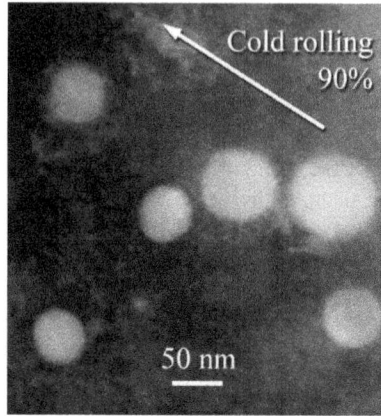

Figure 10.
Bcc Nb precipitates in a Zr—2.5Nb alloy subjected to cold rolling up to 90% [4]. Although the bcc Nb is softer than the hcp Zr in terms of shear modulus, the bcc Nb precipitates in the hcp Zr matrix are actually nonshearable, strong obstacles against gliding dislocations.

precipitates as the slip plane of precipitate interior is not parallel to the slip plane of the matrix.

The hcp Zr matrix of the Zr—2.5Nb alloy contained Nb ~0.5 at.%, which is greater than the solubility (see **Figure 10** of the previous chapter). The excess Nb atoms formed nanoprecipitates when the alloy is subjected to high-energy particle irradiation. Unlike the bcc Nb precipitates whose α is 0.8–1, the α of the Nb nanoprecipitates produced by irradiation was estimated to be ~0.2 or less. Their obstacle strength α is plotted in **Figure 9** as a function of damage level (displacement per atom: dpa). This analysis is based on an assumption that the irradiation-induced hardening occurred solely due to nanoprecipitates. In reality, however, the irradiated samples may also have contained defect clusters such as dislocation loops at high density. This assumption yields an overestimation of the α of nanoprecipitates; nevertheless, the obtained α was extremely small, indicating that the Nb nanoclusters are weak obstacles. The origin of such small α of nanoprecipitates is presumably attributable to the structural change of precipitates

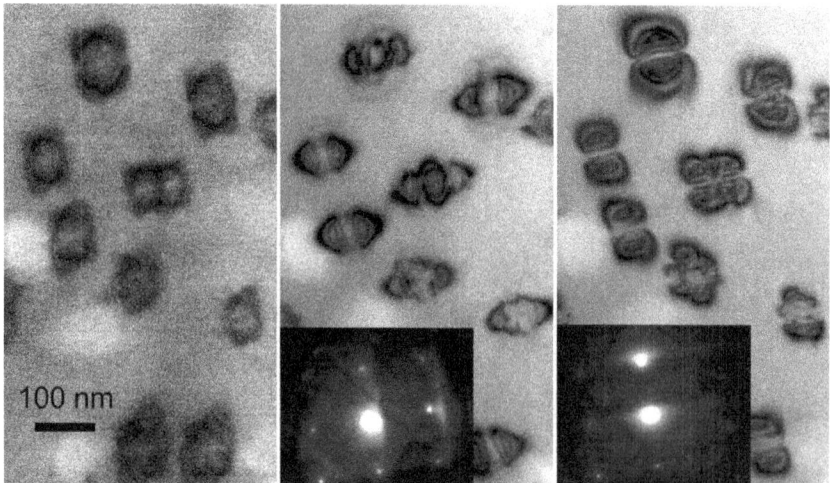

Figure 11.
Coherent fcc Co precipitates embedded in fcc Cu matrix of a Cu-3 wt.% Co alloy [23]. The strain contrast around the precipitate particles in undeformed samples is lost in deformed samples.

42

during precipitation described in the previous chapter. This hypothesis is still open for further investigation.

Soft precipitates can become non-shearable obstacles against dislocations due to the effect of crystallography. Likewise, hard precipitates can become shearable if crystallography allows, i.e., when their slip plane is parallel to that of the matrix. An example shown here is a coherent fcc Co precipitate particle embedded in fcc Cu matrix [23] (**Figure 11**). The shear modulus of the fcc Co is about two times greater than that of the fcc Cu [24]; nevertheless, the Co precipitates are actually shearable (**Figure 12**). It still remains unclear how much hard particles are shearable. It appears that this process occurs only in a limited circumstance. The Co particles were sheared only when interacted with Shockley partial dislocations having the same Burgers vector, gliding on adjacent {111} planes, forming a twin band. Otherwise, dislocations bypassed the Co precipitates via the Hirsch mechanism [25] (**Figure 13**). The Hirsch mechanism is similar to the Orowan mechanism but distinct in terms of the type of dislocation loop remained after the interaction.

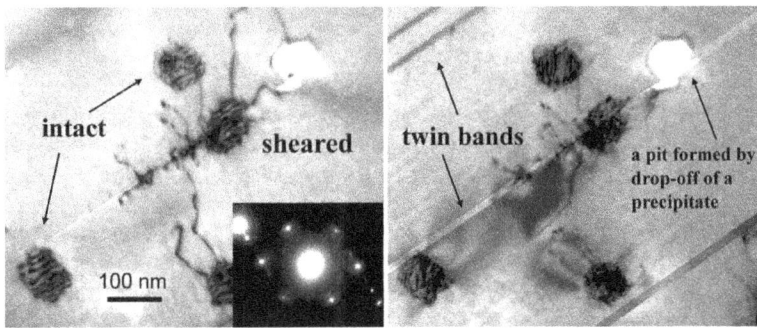

Figure 12.
Cutting of strong obstacles by dislocations: fcc Co precipitates in fcc Cu matrix [25]. The shear modulus of the fcc Co is two times larger than the fcc Cu [24].

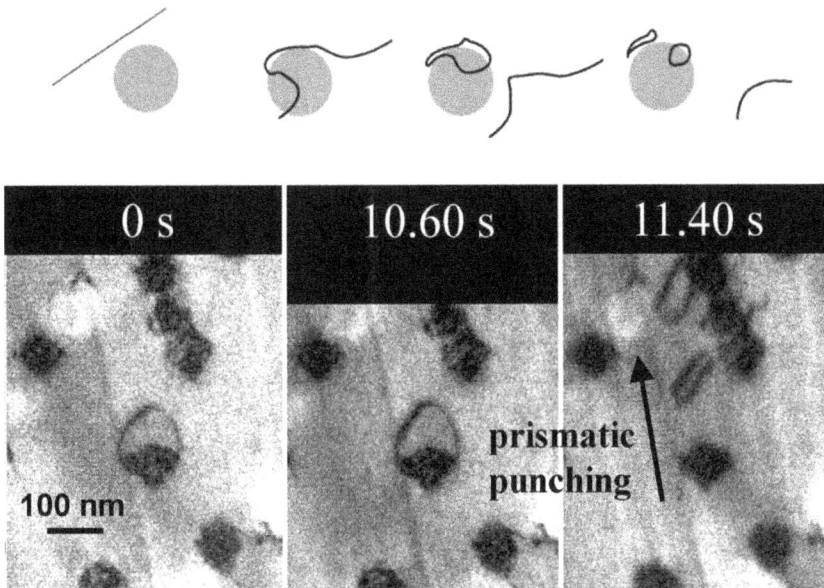

Figure 13.
The Hirsch mechanism [26] observed by TEM in situ straining experiments: fcc Co precipitates in fcc Cu matrix [25].

The Orowan loop is a shear dislocation loop whose Burgers vector is parallel to the loop plane, whereas the Hirsch mechanism produces a prismatic loop [26] whose Burgers vector is not parallel. When the Burgers vector is perpendicular to the loop plane, the prismatic often exhibit one-dimensional back and forth motion along the Burgers vector [27]. The Hirsch mechanism is frequently observed in TEM in situ straining experiments using thin foil specimens, which have a less constraint for deformation in the thickness direction (i.e., so-called plane stress condition). In such a thin foil geometry, screw dislocations that compensate the out-of-plane shear displacements are dominant over edge dislocations [28]. Screw dislocations exhibit cross slip (slip transfer from the slip plane to another slip plane) on non-shearable obstacles. The Hirsch mechanism is induced by the cross slip of screw dislocations on the surface of obstacles [25]. Hence, the Hirsch mechanism is probably dominant over the Orowan mechanism in the deformation of thin foil samples.

4. Precipitation hardening due to solute clusters

Precipitation hardening is a key research subject not only for developing new, strong materials but also for estimating the engineering lifetime of existing materials. For instance, engineering lifetime of reactor pressure vessels (RPVs) of light-water nuclear reactors is determined by embrittlement due to precipitation of minor alloying elements such as Cu, Ni, Mn, and Si rather than accumulation of irradiation damages. Since the RPVs are practically non-replaceable due to economic reasons, their engineering lifetime determines the useful lifetime of entire power plants. Establishing a predictive model of material embrittlement (loss of ductility) is a long-standing challenge in fundamental physical metallurgy. Although the theory of dislocations is well established for quantitatively describing the strength of materials, the dislocation theory is incapable of directly describing the ductility. Hence, the loss of ductility has often been indirectly scaled by the degree of hardening, based on a generally accepted empirical rule that stronger materials exhibit less ductility. The size of irradiation-induced precipitates in the RPV steels is typically a few nm. In the early stage of precipitation, they may be solute clusters rather than second-phase particles crystallographically distinct from the matrix.

In order to evaluate hardening due to solute clusters, the Orowan model needs a modification as follows. The simulation by Foreman and Makin was performed not only on strong obstacles but also on weak obstacles. In **Figure 6** empirical results of their simulations are plotted as a function of bowing angles (θ), together with outputs of two theoretical models: one is the Orowan model with a square lattice arrangement, and the other is a model obtained based on Friedel's geometry consideration (a.k.a the Friedel's statistics) [29, 30] similar to **Figure 8**. In the Friedel's concept the angle (β) of **Figure 14** is assumed to be very small, i.e., L < <R, which is a realistic approximation for weak obstacles. In fact, the simulation results were in good agreement with this model at bowing angles of greater than $\sim$100°. The formula based on the Friedel's approximation is obtained as follows [29, 30]:

$$F = 2T\sin\beta = \tau b(2\beta R) \rightarrow 2T\beta \approx 2\tau b\beta R \rightarrow R = \frac{T}{\tau b} \qquad (13)$$

$$\sin\beta = \frac{L}{R} \rightarrow \beta \approx \frac{L}{R} \qquad (14)$$

$$\tan\left(\frac{\beta}{2}\right) \approx \frac{h}{L} \rightarrow \frac{\beta}{2} \approx \frac{h}{L} \rightarrow h = \frac{L^2}{2R} \qquad (15)$$

$$\Lambda^2 \approx Lh \tag{16}$$

From Eqs. (13) and (15), (16), we obtain.

$$\Lambda^2 = \frac{L^3}{2R} = \frac{\tau b L^3}{2T} \rightarrow L = \left(\frac{2T\Lambda^2}{\tau b}\right)^{1/3} = \left(\frac{\mu b \Lambda^2}{\tau}\right)^{1/3} \tag{17}$$

From Eqs. (4) and (17), the following relationship is obtained [12]:

$$L = \frac{\mu b}{\tau} \cos(\theta/2) = \left(\frac{\mu b \Lambda^2}{\tau}\right)^{1/3}$$

$$\left(\frac{\mu b}{\tau}\right)^3 [\cos(\theta/2)]^3 = \frac{\mu b \Lambda^2}{\tau}$$

$$\tau^2 = \left(\frac{\mu b}{\Lambda}\right)^2 [\cos(\theta/2)]^3$$

$$\tau = \frac{\mu b}{\Lambda} [\cos(\theta/2)]^{3/2} \tag{18}$$

By replacing Λ with L, this formula is generalized as follows [12]:

$$\tau = \frac{\mu b}{L} [\cos(\theta/2)]^{3/2} \; \theta \geq 100° \tag{19}$$

In practice, however, applying Eq. (19) to the analysis of weak obstacles is not straightforward; it is difficult to evaluate how much weak the obstacles are. The Russell-Brown model [22] is an alternative model, more practically useful than the previous model for this purpose (**Figure 15**). This model was originally developed for Cu precipitates in Fe—Cu steels; the crystal structure of which is not fcc but bcc in the early stage of precipitation. In this model the obstacle strength is scaled by the ratio of the energy of dislocation segments in precipitates and in matrix. The energy of dislocations is dependent on the shear modulus. The shear modulus of fcc Cu is lower than bcc Fe. According to the results of ab initio calculations, the shear modulus of bcc Cu is even smaller. The energy of dislocation segment inside the Cu precipitate is lower than that in the matrix Fe.

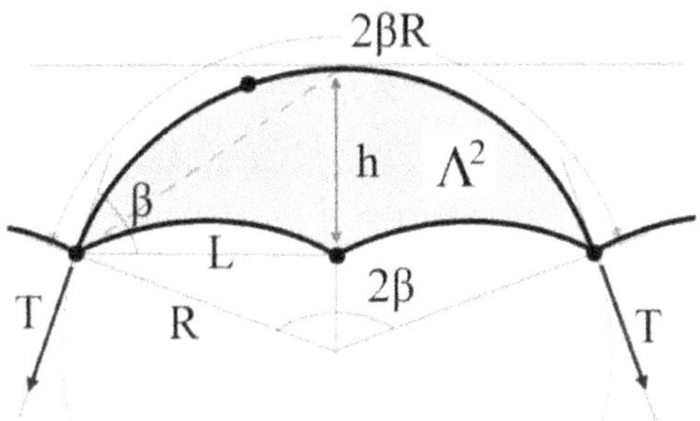

Figure 14.
*Friedel's geometry consideration of dislocation-obstacle interactions [29, 30]. This approximation assumes L < <R, i.e., 2β 0, whereas the other previously discussed in **Figure 8** assumes L = R.*

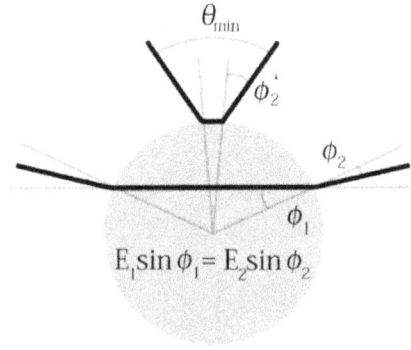

$$E_1 \sin \phi_1 = E_2 \sin \phi_2$$

Figure 15.
The Russell-Brown model for weak obstacles [23]. This model scales the obstacle strength of soft precipitates by the ratio of shear modulus between precipitates and matrix.

In accordance with the aforementioned empirical knowledge obtained from the simulation of Foreman and Makin, on the applicability limit of the Orowan model in terms of dislocation bowing angle, when $\sin^{-1}(E_1/E_2) \leq 50°$,

$$\tau = 0.8\frac{\mu b}{L}\left[1 - \left(\frac{E_1}{E_2}\right)^2\right]^{1/2}, \sigma = 0.8\frac{M\mu b}{L}\left[1 - \left(\frac{E_1}{E_2}\right)^2\right]^{1/2}. \qquad (20)$$

When $\sin^{-1}(E_1/E_2) \geq 50°$,

$$\tau = \frac{\mu b}{L}\left[1 - \left(\frac{E_1}{E_2}\right)^2\right]^{3/4}, \sigma = \frac{M\mu b}{L}\left[1 - \left(\frac{E_1}{E_2}\right)^2\right]^{3/4} \qquad (21)$$

The energy ratio is given as follows:

$$\frac{E_1}{E_2} = \frac{E_1^\infty \log \frac{r}{r_0}}{E_2^\infty \log \frac{R}{r_0}} + \frac{\log \frac{R}{r}}{\log \frac{R}{r_0}} \qquad (22)$$

where E_1^∞ and E_2^∞ refer to the energy per unit length of a dislocation in infinite media, r is the radius of precipitates, and r_0 and R are the inner and outer cutoff radius (they adopted $R = 10^3 r_0$). Since the energy of dislocation is proportional to the shear modulus, the ratio of energy E_1^∞/E_2^∞ is equal to the ratio of shear modulus G_1/G_2 for a screw dislocation, and $\frac{G_1(1-\nu_2)}{G_2(1-\nu_1)}$ for an edge dislocation (ν: the Poisson's ratio)—in the case of bcc Cu precipitates embedded in bcc Fe matrix, 0.59 for the former and 0.64 for the latter.

The Russell-Brown model indicates that, even in the case where precipitates are softer than the matrix, they become weak obstacles only when their shear modulus is slightly smaller than the matrix. Two extreme conditions, $E_1^\infty = E_2^\infty$ and $E_1^\infty = 0$, correspond to the situations of "no obstacle" and "the strongest obstacle," respectively. The latter may be consistent with our empirical knowledge that voids are the strongest obstacles, obtained from experiments [31] and from MD simulations [32, 33]. Although the Russell-Brown model indicates that those which are as soft as the matrix are weak obstacles, as mentioned earlier, such precipitates can also become ideal Orowan-type strong obstacles if the slip plane inside the precipitates is not parallel to that of the matrix [3–4]. From this respect, the situation where precipitate particles become weak obstacles against dislocations may be rather rare. The Russell-Brown model has been applied to the analysis of irradiation-induced hardening of RPV steels due to not only Cu precipitates but also Ni—Si—Mn precipitates [34–37], though it remains unclear whether the Ni—Si—Mn precipitates are softer than the matrix. If they were harder than the matrix ($E_1^\infty > E_2^\infty$), it may not be mathematically valid to apply the Russell-Brown model to them, regardless of whether they are shearable obstacles. Furthermore, the effect of crystal structure change during precipitation has not been considered thus far.

Acknowledgements

The author was supported by the MEXT Grant-in-Aid for Young Scientists (A) (22686058), by Japan Society for the Promotion of Science (JSPS) KAKENHI (#16K06767) and by The Iron & Steel Institute of Japan (ISIJ) the 23rd and the 26th Research Promotion Grants. This review article is based on the author's previous researches partly sponsored by the Ministry of Education, Culture, Sports, Science and Technology (MEXT) of Japan, under the Strategic Promotion Program for Basic Nuclear Researches entitled "Study on hydrogenation and radiation effects in advanced nuclear fuel cladding materials" and "Study of degradation mechanism of stainless steel weld-overlay cladding of nuclear reactor pressure vessels" and a program entitled "R&D of nuclear fuel cladding materials and their environmental degradations for the development of safety standards" entrusted to Tohoku University by the MEXT. Those researches were also supported in part by the Collaborative Research Programs of "the Oarai Center" and "the Cooperative Research & Development Center for Advanced Materials" of the Institute for Materials Research, Tohoku University, and of the Research Institute for Applied Mechanics, Kyushu University; by Advanced Characterization Nanotechnology Platform, Nanotechnology Platform Program of the MEXT, Japan at the Research Center for Ultra-High Voltage Electron Microscopy in Osaka University and at the Ultramicroscopy Research Center in Kyushu University; and by the Joint Usage/ Research Program on Zero-Emission Energy Research, Institute of Advanced Energy, Kyoto University (ZE27C-07, ZE28C-09, ZE29C-11, and ZE30C-01).

Conflict of interest

The author declares no conflicts of interest directly relevant to the content of this article.

Author details

Yoshitaka Matsukawa[1,2]

1 Faculty of Advanced Science and Technology, Division of Materials Science, Kumamoto University, Kumamoto, Japan

2 Institute for Materials Research, Tohoku University, Sendai, Miyagi, Japan

*Address all correspondence to: ym2@msre.kumamoto-u.ac.jp

IntechOpen

References

[1] Orowan E. Discussion. In: Symposium on Internal Stresses in Metals and Alloys, Institute of Metals; Monograph and Report Series 5. London: Institute of Metals; 1948. pp. 451-453

[2] Ardell AJ. Precipitation hardening. Materials Transactions. 1985;**16A**: 2131-2165

[3] Matsukawa Y, Yang HL, Saito K, Murakami Y, Maruyama T, Iwai T, et al. The effect of crystallographic mismatch on the obstacle strength of second phase precipitate particles in dispersion strengthening: bcc Nb particles and nanometric Nb clusters embedded in hcp Zr. Acta Materialia. 2016;**102**: 323-332

[4] Matsukawa Y, Okuma I, Muta H, Shinohara Y, Suzue R, Yang HL, et al. Crystallographic analysis on atomic-plane parallelisms between bcc precipitates and hcp matrix in recrystallized Zr–2.5Nb alloys. Acta Materialia. 2017;**126**:86-101

[5] Rice RM, Zinkle SJ. Temperature dependence of the radiation damage microstructure in V ± 4Cr ± 4Ti neutron irradiated to low dose. Journal of Nuclear Materials. 1998;**258–263**: 1414-1419

[6] Zinkle SJ, Matsukawa Y. Observation and analysis of defect cluster production and interactions with dislocations. Journal of Nuclear Materials. 2004; **329–333**:88-96

[7] Hashimoto N, Byun TS, Farrell K, Zinkle SJ. Deformation microstructure of neutron-irradiated pure polycrystalline metals. Journal of Nuclear Materials. 2004;**329–333**: 947-952

[8] Armstrong RW. 60 years of Hall-Petch: Past to present nano-scale

connections. Materials Transactions. 2014;**55**:2-12

[9] Shen JH, Li YL, Wei Q. Statistic derivation of Taylor factors for polycrystalline metals with application to pure magnesium. Materials Science and Engineering A. 2013;**582**:270-275

[10] Leguey T, Baluc N, Schäublin R, Victoria M. Structure-mechanics relationships in proton irradiated pure titanium. Journal of Nuclear Materials. 2002;**307–311**:696-700

[11] Conrad H. The rate controlling mechanism during yielding and flow of titanium at temperatures below 0.4 T_m. Acta Metallurgica. 1966;**14**:1631-1633

[12] Foreman AJE, Makin MJ. Dislocation movement through random arrays of obstacles. Philosophical Magazine. 1966;**15**:911-924

[13] Kocks UF. A statistical theory of flow stress and work hardening. Philosophical Magazine. 1966;**13**: 541-546

[14] Kocks UF. On the spacing of dispersed obstacles. Acta Metallurgica. 1966;**14**:1629-1631

[15] Kocks UF. Statistical treatment of penetrable obstacles. Canadian Journal of Physics. 1967;**45**:737-755

[16] Ashby MF. The theory of the critical shear stress and work hardening of dispersion-hardened crystals. In: Ansell GS et al., editors. Oxide Dispersion Strengthening. New York: Gordon and Breach; 1968. pp. 143-205

[17] Seeger AK. 2nd UN Conference on Peaceful Uses of Atomic Energy. Vol. 6. United Nations, New York; 1958. p. 250

[18] Delesse A. Ueber die mineralogische und chemische Beschaffenheit der

Gesteine der Vogesen. Journal für Praktische Chemie. 1948;**43**:417-448

[19] DeHoff RT, Rhines FN. Quantitative Microscopy. New York: McGraw-Hill; 1968. pp. 21-23, 139-140

[20] Bacon DJ, Osetsky YN, Rodney D. Dislocation-obstacle interactions at the atomic level. In: Hirth JP, Kubin L, editors. Dislocations in Solids. Amsterdam: Elsevier; 2009. pp. 75, 85-136, 228-237

[21] Integrity of Reactor Pressure Vessels in Nuclear Power Plants. Assessment of Irradiation Embrittlement Effects in Reactor Pressure-Vessel Steels. IAEA Nuclear Energy Series No. NP-T-3.11. Vienna: International Atomic Energy Agency; 2009. pp. 63-63

[22] Russell KC, Brown LM. A dispersion strengthening model based on differing elastic moduli applied to the iron-copper system. Acta Metallurgica. 1972;**20**: 969-974

[23] Matsukawa Y, Liu GS. In-situ TEM study on elastic interaction between a prismatic loop and a gliding dislocation. Journal of Nuclear Materials. 2012;**425**: 54-59

[24] Shim J-H, Voigt H-JL, Wirth BD. Temperature dependent dislocation bypass mechanism for coherent precipitates in Cu-Co alloys. Acta Materialia. 2016;**110**:276-282

[25] Matsukawa Y. Unpublished data

[26] Hatano T. Dynamics of a dislocation bypassing an impenetrable precipitate: The Hirsch mechanism revisited. Physical Review B. 2006;**74**:020102(R)

[27] Matsukawa Y, Zinkle SJ. One-dimensional fast migration of vacancy clusters in metals. Science. 2007;**318**: 959-962

[28] Matsukawa Y, Osetsky YN, Stoller RE, Zinkle SJ. Mechanisms of stacking fault tetrahedra destruction by gliding dislocations in quenched gold. Philosophical Magazine. 2008;**88**: 581-597

[29] Friedel J. Dislocations. Oxford: Paragon Press; 1964. pp. 220-228, 371-382

[30] Friedel J. On the elastic limit of crystals. In: Electron Microscopy and Strength of Crystals. Proc. 1st Barkeley Int'l Mater. Conf. New York: Interscience Publishers (A Division of John Wiley & Sons); 1963. pp. 605-649

[31] Lucas GE. The evolution of mechanical property change in irradiated austenitic stainless steels. Journal of Nuclear Materials. 1993;**206**: 287-305

[32] Bacon DJ, Osetsky YN, Rodney D. Dislocation-obstacle interactions at the atomic level. In: Hirth JP, Kubin L, editors. Dislocations in Solids. Vol. 15. Amsterdam: Elsevier; 2009. pp. 1-160

[33] Hatano T, Matsui H. Molecular dynamics investigation of dislocation pinning by a nanovoid in copper. Physical Review B. 2005;**72**:094105

[34] Takeuchi T, Kuramoto A, Kameda J, Toyama T, Nagai Y, Hasegawa M, et al. Effects of chemical composition and dose on microstructure evolution and hardening of neutron-irradiated reactor pressure vessel steels. Journal of Nuclear Materials. 2010;**402**:93-101

[35] Kuramoto A, Toyama T, Takeuchi T, Nagai a Y, Hasegawa M, Yoshiie T, et al. Post-irradiation annealing behavior of microstructure and hardening of a reactor pressure vessel steel studied by positron annihilation and atom probe tomography. Journal of Nuclear Materials. 2012;**425**:65-70

[36] Kuramoto A, Toyama T, Nagai Y, Inoue K, Nozawa Y, Hasegawa M, et al. Microstructural changes in a Russian-type reactor weld material after neutron irradiation, post-irradiation annealing and re-irradiation studied by atom probe tomography and positron annihilation spectroscopy. Acta Materialia. 2013;**61**:5236-5246

[37] Shimodaira M, Toyama T, Yoshida K, Inoue K, Ebisawa N, Tomura K, et al. Contribution of irradiation-induced defects to hardening of a low copper reactor pressure vessel steel. Acta Materialia. 2018;**155**:402-409

Chapter 4

Covariant Dynamical Theory of X-Ray Diffraction

Arthur Dyshekov and Yurii Khapachev

Abstract

The proposed nonstandard diffraction theory is constructed directly from the Maxwell equations for the crystalline medium in the X-ray wavelength range. Analysis of Maxwell's equations for dynamic diffraction is possible using the method of multiple scales which is modified to the vector character of the problem. In this case, the small parameter of the expansion is the Fourier component of the polarizability of the crystal. The second-order wave equation is analyzed without any assumptions about the possibility of the interaction between the refracted and scattered waves which automatically leads to the dynamic character of the scattering. The unified consideration of different geometrical schemes of diffraction including grazing geometry is possible. This is due to the construction of a unified wave field in the crystal and obtaining the field amplitudes according to the boundary conditions. The proposed theory allows generalization to the case of an imperfect crystal. Thus, a unified approach to account for deformations and other crystal structure disturbances in all diffraction schemes is implemented. The determination of a unified wave field without separation of the refracted and scattered waves is of the greatest importance in the analysis of secondary processes.

Keywords: X-ray diffraction, dynamic theory, imperfect crystal, perturbation theory, the method of multiple scales, deformation, extinction length, boundary conditions, reflection coefficient

1. Introduction

It is possible to separate several fundamental approaches in the theory of dynamical X-ray scattering in the crystal [1, 2].

The Darwin theory [1] is based on the Bragg model for the crystal considered as a family of parallel crystal planes. The X-ray wave reflection is considered as a result of successive transmission and multiple reflections from planes. In this case determination of the diffracted wave amplitude is reduced to the solution of recurrent relations between the amplitudes of transmitted and scattered waves in passing through the specified atomic plane. In essence, the Darwin theory represents a direct extrapolation of the optical task of propagation of light in a layer continuum to the case of wavelengths of the X-ray range.

The Evald-Laue theory [2] was the next stage in the development of theoretical approximations about the character of X-ray wave propagation in the crystal under dynamic scattering. The model approximations of the X-ray crystal interaction were formulated in the framework of this theory, meaning that the X-ray wavelength is comparable to the interatomic distances.

Therefore, the standard continual approximation for electrodynamics of continua proves to be unacceptable, and the scattering from individual charges should be taken into consideration. As is known, taking this into account results in the formalism of 3D periodic dielectric permeability $\varepsilon(\mathbf{r})$ or polarizability $\chi(\mathbf{r})$ with the lattice spacings of the crystal. The Evald-Laue theory proceeds from the conception of a uniform wave field that appears in the crystal under dynamic scattering. In the two-wave approximation, the wave field is a superposition of refracted and diffracted waves. The determination of field amplitudes is reduced to the solution of a certain dispersion equation following from the fundamental equations of the theory.

In spite of a series of unconditional advantages in the interpretation and theoretical prediction of experimental results on dynamic X-ray scattering in crystals, both the Evald-Laue theory and, substantially, the Darwin theory have the principal limitation that they describe the dynamical diffraction in perfect crystals only.

The necessity of taking into accounts the different deviations from ideal periodicity in the crystal and, first of all, deformations resulted in the creation of the generalized theory developed by Takagi and Taupin [2]. This is based on the approximation of a wave field in the form of superposition of the transmitted and diffracted waves with slowly varied amplitudes depending on the coordinates that leads to the Takagi-Taupin equations to the system of differential equations relative to the field amplitudes. This formalism gives the possibility to describe the dynamical diffraction in a distorted crystal since the distortions of ideal periodicity can be taken into account in the explicit form in the wave field approximation. Correspondingly, the Takagi-Taupin equations become the system of differential equations with variable coefficients.

It is important that the Takagi-Taupin system is shortened and the coordinate second derivatives of field amplitudes are neglected in it. On the one hand, this significantly facilitates the theoretical consideration and makes observable the solution of a series of diffraction problems in standard diffraction geometries when this simplification proves to be justified. On the other hand, the Takagi-Taupin equations become inapplicable under the conditions, for example, of grazing diffraction geometry; then, it is necessary to solve the third- or even fourth-order differential equations [3].

The principal disadvantage of the procedure of equation shortening is related to the impossibility to state correctly the boundary conditions at the crystal-vacuum interface for field amplitudes. Instead of the known classical continuity conditions for tangential components of electric and magnetic fields, the boundary conditions of the type of setting of the normal components of field amplitudes on the crystal surface become vaguely clear but not in line with the Maxwell equations. Of cause the solutions of these boundary problems prove to be applicable only for rather large (more correctly significantly exceeding the angle of full external reflection) angles of radiation incidence and yield.

As a result the theory faces the difficulties related to the necessity to solve the third- or fourth-order equations which become virtually overwhelming for the case of the crystal with lattice deformation when the diffraction schemes of the type of sliding diffraction are considered.

At the same time, the Maxwell equations are the first-order equations or the second-order ones in the case of a single wave equation when passing, for example, to an electric field. Namely, this structure of equations agrees with the mentioned classical boundary conditions. This means that the requirement of taking correctly into account the boundary conditions in any theoretical diffraction scheme leads virtually unambiguously to the known structure of wave equation following from the Maxwell equation.

Thus, to overcome the aforementioned difficulties, it is necessary to create the theory directly based on the Maxwell equations using model approximations of crystal polarizability in the X-ray wavelength range.

2. A covariant dynamical theory of X-ray scattering in perfect crystals

In this section we follow the original papers [4–7].

2.1 Physical diffraction model and basic equation

The model approximations for crystal and X-rays propagating in it are reduced to the following.

The plane monochromatic wave falls from vacuum onto the crystal. The crystal-vacuum interface is considered as a geometrical one so that the classical boundary conditions for optics prove to be applicable. The uniform wave field in the crystal is described by the fundamental Maxwell equations supplemented by the constitutive equation $D = \varepsilon(r)E$ with 3D periodic dielectric permeability $\varepsilon(r)$ or polarizability $\chi(r)$ with the lattice spacing. Thereby, the linearity and isotropy of continuum in the X-ray wavelength range and the locality of connection between D and E are suggested. The possible time dependence is not taken into account; therefore, the possibility of incoherent (in the sense of the change in the radiation frequency) scattering is eliminated. As is seen, the indicated suggestions properly correspond to the model underlying the Evald-Laue theory if the functional form $\varepsilon(r)$ is not specified.

The Maxwell equations, as the equations of electromagnetic waves in dielectric, when dispersion is absent, can be written in the form (designations here and below are standard):

$$\text{rot}E = -\frac{1}{c}\frac{\partial H}{\partial t}; \tag{1}$$

$$\text{div}H = 0; \tag{2}$$

$$\text{rot}H = \frac{1}{c}\frac{\partial D}{\partial t} = \frac{1}{c}\frac{\partial}{\partial t}(1+\chi(r'))E = \frac{1}{c}(1+\chi(r'))\frac{\partial E}{\partial t}; \tag{3}$$

$$\text{div}D = 0 \tag{4}$$

These equations are supplemented by the constitutive relations:

$$D = \varepsilon(r')E = (1+\chi(r'))E; \quad B = \mu H = H \tag{5}$$

As usual, we will consider that the time dependence of E and H is harmonic:

$$E(r',t) = E(r')\exp(-i2\pi\nu t); \quad H(r',t) = H(r')\exp(-i2\pi\nu t).$$

We have the system

$$\text{rot}E = -\frac{1}{c}\frac{\partial H}{\partial t} = ikH;$$
$$\text{rot}H = \frac{1}{c}(1+\chi(r'))\frac{\partial E}{\partial t} = -ik(1+\chi(r'))E. \tag{6}$$

For economy we use here and below factor 2π into k, $2\pi k \rightarrow k$. We come from system (6) to the basic equation in the standard way:

$$\text{rotrot}\mathbf{E} - k^2(1 + \chi(\mathbf{r}'))\mathbf{E} = 0. \tag{7}$$

In the present section, we will consider the case of a perfect crystal. This will allow us to compare the conclusions of the proposed formalism to the known theoretical results.

We choose the crystal model $\chi(\mathbf{r}')$ in the form

$$\chi(\mathbf{r}') = \chi_0 + \chi_H \exp(i2\pi\mathbf{H}\mathbf{r}') + \chi_{\overline{H}} \exp(-i2\pi\mathbf{H}\mathbf{r}'). \tag{8}$$

Below, we will bring 2π into $\mathbf{H}$ as for k, $2\pi\mathbf{H} \rightarrow \mathbf{H}$. Then, we have for Eq. (7)

$$\text{rotrot}\mathbf{E}(\mathbf{r}') - k^2\left(1 + \chi_0 + \chi_H \exp(i\mathbf{H}\mathbf{r}') + \chi_{\overline{H}} \exp(-i\mathbf{H}\mathbf{r}')\right)\mathbf{E}(\mathbf{r}') = 0. \tag{9}$$

To take correctly into account the contribution of different terms into Eq. (9) when using the methods of perturbation theory, it is necessary to bring it to the dimensionless form. This procedure assumes the choice of a certain characteristic spatial scale of the problem. Apparently, this is the parameter determining the reciprocal lattice in our problem, namely, the modulus of the reciprocal lattice vector:

$$\frac{\mathbf{H}}{H} = \mathbf{h}; \quad \mathbf{H}\mathbf{r}' = \mathbf{h}\mathbf{r}; \quad \mathbf{r} = H\mathbf{r}'.$$

$$\text{rotrot}\mathbf{E}(\mathbf{r}) - \kappa^2\left(1 + \chi_0 + \chi_H \exp(i\mathbf{h}\mathbf{r}) + \chi_{\overline{H}} \exp(-i\mathbf{h}\mathbf{r})\right)\mathbf{E}(\mathbf{r}) = 0; \tag{10}$$

$$\kappa = \frac{k}{H}.$$

Eq. (10) cannot be exactly solved. Correspondingly, it is necessary to use some method of approximate solution. There are two main ways to analyze Eq. (10). In the first case, taking into account that $\chi(\mathbf{r}') \ll 1$ Eq. (10) is presented in the form of an inhomogeneous equation, the right side of which is considered as a small perturbation specifying the field of the incident wave:

$$\text{rotrot}\mathbf{E}(\mathbf{r}') - \kappa^2\mathbf{E}(\mathbf{r}') = \chi(\mathbf{r}')\mathbf{E}(\mathbf{r}').$$

The approximate solution is searched in the space beyond the scattering crystal at large distances from it in the form of the first term of a series of the Born expansion. The applicability of this approach is limited by the smallness of the scattering cross section as compared to the geometrical area of the crystal section. As is known this approach results in the kinematical theory [8].

In the second case, two variants of the dynamical theory are developed from Eq. (10). For the first variant, the solution of Eq. (10) is sought in the form of Bloch wave represented in the form of an infinite series of plane waves with wave vectors corresponding to the refracted wave and diffracted waves in the crystal. This Bloch wave is interpreted as a multiwave solution of the dynamical theory. As a result, the use of expansion of $\chi(\mathbf{r}')$ in a Fourier series leads to an infinite system of the fundamental equations of the algebraic Evald-Laue type.

Since it is not possible to solve an infinite system, one has to be limited as a rule by two equations, that is, by the two-wave approximation.

For the second variant of the theory, the solution of Eq. (10) is presented in the form of plane waves with the slowly varying amplitude. As a result of the two-wave approximation, we obtain the Takagi-Taupin equations which can be interpreted as the recurrent Darwin relations written in the differential form [2].

We propose here to use a new approach to the analysis of Eq. (10). The physical justification of the proposed method indicates the choice of the perturbation parameter and is as follows. The propagation of an X-ray wave in the crystal unaccompanied by the appearance of diffracted beams is adequately described by the uniform wave equation with $\varepsilon = 1 + \chi_0$. This corresponds to the propagation of an X-ray wave in the crystal as a continuum with the refraction factor with respect to the usual laws of optics. This situation can be considered typical. On the contrary the appearance of diffracted beams requires that the definite geometrical conditions be fulfilled for the wave vectors and reciprocal lattice vector. Apparently, χ_H is responsible for this cardinal change in the picture of a wave field in the crystal.

Thus, in spite of the fact that all quantities $\chi_0, \chi_H, \chi_{\bar{H}} \sim 10^{-5} \div 10^{-6}$, namely, χ_H, should be chosen as the perturbation parameter.

2.2 Direct expansion and geometric diffraction conditions

We use the simplest perturbation method, direct expansion, in χ_H:

$$E(r) = E_0(r) + \chi_H E_1(r) + \chi_H^2 E_2(r) + \dots$$

We will restrict ourselves to the first-order expansion. We have

$$\text{rotrot}E_0 + \chi_H \text{rotrot}E_1 + \dots$$
$$- \left(\kappa^2(1+\chi_0) + \kappa^2 \chi_H \left(\exp(i\mathbf{hr}) + \frac{\chi_{\bar{H}}}{\chi_H} \exp(-i\mathbf{hr}) \right) \right) (E_0 + \chi_H E_1 + \dots) = 0 \quad (11)$$

The zeroth approximation corresponds to zero power of the perturbation parameter χ_H:

$$\text{rotrot}E_0 - \kappa_0^2 E_0 = 0;$$
$$\kappa_0^2 = \kappa^2(1+\chi_0).$$

We select the solution of this vector wave equation in the form of superposition of two plane waves:

$$E_0(r) = E_{01}(r) \exp(i\boldsymbol{\kappa}_0 r) + E_{02}(r) \exp(-i\boldsymbol{\kappa}_0 r);$$
$$(\boldsymbol{\kappa}_0 E_i) = 0$$

The reasons for this selection are the following. The zeroth approximation corresponds to the propagation of two plane transverse waves in opposite directions in the continuum with $\chi = \text{const}$. This is a singular analog of the total field in the crystal for the case of an empty lattice. The propagation directions and the wave amplitudes remain indeterminate and are specified below by the boundary conditions at the vacuum-crystal interface.

The first approximation is obtained when all terms in Eq. (11) proportional to the first power of χ_H are zero:

$$\text{rotrot}E_1 - \kappa_0^2 E_1 = \kappa^2 \left(\exp(i\mathbf{hr}) + \frac{\chi_{\bar{H}}}{\chi_H} \exp(-i\mathbf{hr}) \right) E_0$$
$$= \kappa^2 \left(\exp(i(\boldsymbol{\kappa}_0 + \mathbf{h})r) + \frac{\chi_{\bar{H}}}{\chi_H} \exp(i(\boldsymbol{\kappa}_0 - \mathbf{h})r) \right) E_{01} + \qquad . \quad (12)$$
$$+ \kappa^2 \left(\exp(-i(\boldsymbol{\kappa}_0 - \mathbf{h})r) + \frac{\chi_{\bar{H}}}{\chi_H} \exp(-i(\boldsymbol{\kappa}_0 + \mathbf{h})r) \right) E_{02}$$

We obtained the inhomogeneous wave equation. According to the perturbation theory, it is necessary to find its particular solution. Since the continuum is uniform with an accuracy of the reciprocal lattice vector $\mathbf{h}$, the desired wave field is delocalized. It means that the particular solution of Eq. (12) must have the form of a plane wave. The particular solution of the equation

$$\text{rotrot} \mathbf{E}_0 - \kappa^0{}_2 \mathbf{E}_0 = \mathbf{A} \exp\left(i\mathbf{q}\mathbf{r}\right)$$

takes the form

$$\mathbf{E} = \frac{\mathbf{A} \exp\left(i\mathbf{q}\mathbf{r}\right)}{q^2 - \kappa_0^2} \tag{13}$$

Eq. (13) is obtained taking into account the condition $\text{div}\mathbf{D} = 0$ from which follows $(\kappa_0 \mathbf{E}_{0i}) = (h\mathbf{E}_{0i}) = 0$ i = 1, 2, that is, the field is strictly transverse when sources are absent.

Now, the solution in the first order of the perturbation theory can be written:

$$\mathbf{E}_1 = \kappa^2 \frac{\mathbf{E}_{01} \exp\left(i(\kappa_0 + \mathbf{h})\mathbf{r}\right)}{\left(\kappa_0 + \mathbf{h}\right)^2 - \kappa_0^2} + \kappa^2 \frac{\chi_{\overline{H}}}{\chi_H} \frac{\mathbf{E}_{01} \exp\left(i(\kappa_0 - \mathbf{h})\mathbf{r}\right)}{\left(\kappa_0 - \mathbf{h}\right)^2 - \kappa_0^2}$$

$$+ \kappa^2 \frac{\mathbf{E}_{02} \exp\left(-i(\kappa_0 - \mathbf{h})\mathbf{r}\right)}{\left(\kappa_0 - \mathbf{h}\right)^2 - \kappa_0^2} + \kappa^2 \frac{\chi_{\overline{H}}}{\chi_H} \frac{\mathbf{E}_{02} \exp\left(-i(\kappa_0 + \mathbf{h})\mathbf{r}\right)}{\left(\kappa_0 + \mathbf{h}\right)^2 - \kappa_0^2} \tag{14}$$

Finally, the direct expansion with an accuracy of χ_H^2 takes the form

$$\mathbf{E}(\mathbf{r}) = \mathbf{E}_0(\mathbf{r}) + \chi_H \mathbf{E}_1(\mathbf{r}) + \dots =$$

$$= \mathbf{E}_{01} \left(\exp\left(i\kappa_0 \mathbf{r}\right) + \kappa^2 \chi_H \frac{\exp\left(i(\kappa_0 + \mathbf{h})\mathbf{r}\right)}{\left(\kappa_0 + \mathbf{h}\right)^2 - \kappa_0^2} + \kappa^2 \chi_{\overline{H}} \frac{\exp\left(i(\kappa_0 - \mathbf{h})\mathbf{r}\right)}{\left(\kappa_0 - \mathbf{h}\right)^2 - \kappa_0^2} \right) +$$

$$\mathbf{E}_{02} \left(\exp\left(-i\kappa_0 \mathbf{r}\right) + \kappa^2 \chi_H \frac{\exp\left(-i(\kappa_0 - \mathbf{h})\mathbf{r}\right)}{\left(\kappa_0 - \mathbf{h}\right)^2 - \kappa_0^2} + \kappa^2 \chi_{\overline{H}} \frac{\exp\left(-i(\kappa_0 + \mathbf{h})\mathbf{r}\right)}{\left(\kappa_0 + \mathbf{h}\right)^2 - \kappa_0^2} \right) + \dots \tag{15}$$

It follows from Eq. (15) that in addition to the direct (κ_0) and inverse ($-\kappa_0$) directions of the plane wave propagation in a continuum, waves in the directions $(\kappa_0 \pm \mathbf{h})$ and $-(\kappa_0 \pm \mathbf{h})$ appear as well. The amplitude of these waves is negligibly small (χ_H times smaller) as compared to the initial one and cannot substantially change the wave field in the crystal. Thus, the refracted (and also possibly reflected) wave with small distortions propagates in the crystal.

However, this position radically changes when any denominator in Eq. (15) approaches zero. In this case $\mathbf{E}_1 \to \infty$, and we cannot consider a small correction to $\mathbf{E}_0$. Then, the direct expansion does not hold, and its modification is required. Apparently, this occurs under condition

$$\left(\kappa_0 \pm \mathbf{h}\right)^2 - \kappa_0^2 \leq \chi_H$$

This condition is well known: it is the Laue condition for X-rays, and therefore there is no need to detail its physical sense. Note only that all geometric constructions following from the Laue condition appear in this case as a natural consequence of validity violation of the direct field expansion in the parameter χ_H.

Thus, the wave field structure principally changes for certain κ_0 values, and new directions of the wave propagation different from the initial one appear, that is, diffraction. We will restrict ourselves here and below to the two-wave approximation when the transmitted and diffracted waves satisfy the Laue equation.

Thus, it is necessary to modify the direct expansion near the κ_0 values for which diffraction is observed. The parametric character of the interaction of continuum with a wave field is the principal moment here which provides the physical (and mathematical) justification for the search of the solution.

There are different methods to modify the direct expansion. All of them are directed to solve one problem: to obtain a so-called uniformly acceptable expansion near the values of parameters interesting for us. The method of multiple scales is most favorable for our investigation [9]. However, method modification is necessary having in mind the vector character of the problem.

2.3 The method of multiple scales

The main idea of the method for the considered problem is the following. The wave field singularities appear for different spatial scales determined by a small parameter of expansion of χ_H. Correspondingly, these singularities can be independently considered in the specified approximation. It is attained mathematically by the transition from one spatially variable ($\mathbf{r}$ in our case) to several ones reflecting different scales of the problems. The fixed number of scales determines the expansion order of the solution. The abovementioned modification is related to the fact that the method of many scales is used for scalar equations; here, it is used for the vector equation.

Thus, we will search for the approximate solution of Eq. (10) for the most interesting case in the area of the Bragg maximum when the Laue condition $\mathbf{\kappa}_0 \pm \mathbf{h} = \mathbf{\kappa}_h$ is fulfilled. We make the substitution $\mathbf{r} \rightarrow \mathbf{r}_0, \mathbf{r}_{1,...} = \mathbf{r}_0, \chi_H \mathbf{r}_0,...$ in Eq. (10), assuming that the field is determined by different spatial scales:

$$\mathbf{E}(\mathbf{r}) = \mathbf{E}(\mathbf{r}_0, \mathbf{r}_1, ...).$$

Hereinafter, we will restrict ourselves to the first order of expansion; correspondingly, we will consider two spatial scales $\mathbf{r}_0$ and $\mathbf{r}_1$.

Then, for rot$\mathbf{E}$ we obtain

$$\text{rot}\mathbf{E} = (\text{rot}_0 + \chi_H \text{rot}_1 + ...)\mathbf{E}$$

that is, operator rot is linear relative to the carried-out substitution. The index of the operator signifies the space in which it operates. Using this property we obtain

$$\text{rotrot} = (\text{rot}_0 + \chi_H \text{rot}_1 + ...)(\text{rot}_0 + \chi_H \text{rot}_1 + ...) =$$

$$= \text{rot}_0 \text{rot}_0 + \chi_H \text{rot}_1 \text{rot}_0 + \chi_H \text{rot}_0 \text{rot}_1 + ...$$

As was indicated above, the interaction of the field with continuum has a parametric character. This means that along with the field expansion it is also necessary to expand the wave vector $\mathbf{\kappa}_0$ in powers of χ_H:

$$\mathbf{E} = \mathbf{E}_0(\mathbf{r}_0, \mathbf{r}_1, ...) + \chi_H \mathbf{E}_1(\mathbf{r}_0, \mathbf{r}_1, ...) + ...;$$

$$\mathbf{\kappa}_0 = \mathbf{\kappa}_{00} + \chi_H \mathbf{\kappa}_{01} + ...; \tag{16}$$

$$\kappa_0^2 = (\mathbf{\kappa}_0, \mathbf{\kappa}_0) = \kappa_0^2 + 2\chi_H(\mathbf{\kappa}_0, \mathbf{\kappa}_{01}) + ... = \kappa_0^2 + \chi_H X; \quad X = 2(\mathbf{\kappa}_0, \mathbf{\kappa}_{01})$$

We substitute all expansions into Eq. (10):

$$(\text{rot}_0 \text{rot}_0 + \chi_H \text{rot}_1 \text{rot}_0 + \chi_H \text{rot}_0 \text{rot}_1 + ...)(\mathbf{E}_0 + \chi_H \mathbf{E}_1 + ...) -$$

$$- \left(\kappa_{00}^2 + \chi_H X + ... + \kappa^2 \chi_H \left(\exp(i\mathbf{h}\mathbf{r}_0) + \frac{\chi_{\bar{H}}}{\chi_H} \exp(-i\mathbf{h}\mathbf{r}_0) \right) \right)(\mathbf{E}_0 + \chi_H \mathbf{E}_1 + ...) = 0 \tag{17}$$

In this case, $\chi(\mathbf{r})$ is presented as a function of the ground spatial scale $\mathbf{r}_0$. The subsequent procedure follows the standard scheme of perturbation methods.

Namely, the initial approximation (unperturbed state) and subsequent ones are obtained when the constants of powers of the perturbation parameter χ_H are sequentially equated. The uniformly available expansion is obtained when additional conditions are imposed elimination of secular (divergent) terms of expansion. In turn, this elimination is due to the expansion of κ_0 and introduction of different scales of the problem.

Let us demonstrate this procedure. As before the zeroth approximation, (the unperturbed equation) has the form of the standard vector wave equation for transverse waves propagating in continuum:

$$\text{rot}_0 \text{rot}_0 \mathbf{E}_0 - \kappa_{00}^2 \mathbf{E}_0 = 0. \tag{18}$$

However, in contrast to the direct expansion, the operator rot acts here only on the single spatial scale $\mathbf{r}_0$. According to this, the solution should search in the form of superposition of transmitted and diffracted waves (two-wave approximation):

$$\mathbf{E}_0 = \mathbf{e}_1 c_1(\mathbf{r}_1) \exp(i\kappa_{00}\mathbf{r}_0) + \mathbf{e}_2 c_2(\mathbf{r}_1) \exp(i\kappa_h \mathbf{r}_0)$$

$$(\kappa_{00}, \mathbf{e}_i) = (\kappa_h, \mathbf{e}_i) = 0. \tag{19}$$

The quantities $c_i(\mathbf{r}_1)$ are related to the other spatial scale and considered as slow variables.

This structure of the wave field supposes strict fulfillment of the diffraction Laue condition $\kappa_0 \pm \mathbf{h} = \kappa_h$ and the presence of reflecting plane. In fact the condition $(\kappa_0 \mathbf{h}) = \pm 1/2$ follows from $(\kappa_0 \pm \mathbf{h})^2 - \kappa_0^2 = 0$ which imposes the limitation only on the component κ_0 directed along $\mathbf{h}$, $\kappa_{0\text{II}}$. The component $\kappa_{0\perp}$ normal to $\mathbf{h}$ is identical for κ_0 and κ_h.

The condition of wave transversity $(\kappa_{00}, \mathbf{e}_i) = (\kappa_h, \mathbf{e}_i) = 0$ defines only the planes orthogonal to the corresponding wave vectors. As is known two cases of polarization are considered, namely, σ polarization when the field amplitude is in the plane orthogonal to the diffraction plane and π polarization when the field amplitude is in the diffraction plane. The case of σ polarization is more favorable for the further consideration being a simpler one.

The following first-order approximation to leads to the inhomogeneous equation:

$$\text{rot}_0 \text{rot}_0 \mathbf{E}_1 - \kappa_{00}^2 \mathbf{E}_1 = - (\text{rot}_1 \text{rot}_0 + \text{rot}_0 \text{rot}_1)\mathbf{E}_0 + X\mathbf{E}_0$$

$$+ \kappa^2 (\exp(i\mathbf{h}\mathbf{r}_0) + \chi_{\bar{H}}/\chi_H \exp(-i\mathbf{h}\mathbf{r}_0))\mathbf{E}_0 \tag{20}$$

Hence

$$\text{rot}_0 \text{rot}_0 \mathbf{E}_1 - \kappa_{00}^2 \mathbf{E}_1 = \left((i2(\nabla_1 c_1, \kappa_{00}) + Xc_1)\mathbf{e}_1 + \kappa^2 \frac{\chi_{\bar{H}}}{\chi_H} c_2 \mathbf{e}_2 \right) \exp(i\kappa_{00}\mathbf{r}_0) +$$

$$+ \left(\kappa^2 c_1 \mathbf{e}_1 + (i2(\nabla_1 c_2, \kappa_h) + Xc_2)\mathbf{e}_2 \right) \exp(i\kappa_h \mathbf{r}_0) + \tag{21}$$

$$+ \kappa^2 \frac{\chi_{\bar{H}}}{\chi_H} \exp(i(\kappa_{00} - \mathbf{h})\mathbf{r}_0) c_1 \mathbf{e}_1 + \kappa^2 \exp(i(\kappa_h + \mathbf{h})\mathbf{r}_0) c_2 \mathbf{e}_2.$$

Operator ∇_1 (gradient) acts here on r1. Eq. (21) is obtained taking into account $(\kappa_{00}\mathbf{e}) = (\kappa_h \mathbf{e}) = 0$ and the additional condition $(\nabla_1 c_1, \mathbf{e}_1) = 0$. This means the following. The quantities $c_i(r_1)$ are considered as the perturbed amplitudes of the corresponding plane waves. The two first terms in the right side of Eq. (21) generate the secular components in expansion which is seen from the abovementioned particular solution. In other words they provide parametric resonance in the system. Consequently, it is necessary to eliminate these terms in order to obtain the uniform approximation near the Laue condition. Then, we obtain the following system of vector equations:

$$\begin{cases} (i2(\nabla_1 c_1, \mathbf{\kappa}_{00}) + X c_1)\mathbf{e}_1 + \kappa^2 \dfrac{\chi_{\overline{H}}}{\chi_H} c_2 \mathbf{e}_2 = 0 \\ (i2(\nabla_1 c_2, \mathbf{\kappa}_h) + X c_2)\mathbf{e}_2 + \kappa^2 c_1 \mathbf{e}_1 = 0 \end{cases} \tag{22}$$

In contrast to the usual scalar system, additional limitations are necessary to solve Eq. (22). The issue is that $\mathbf{e}_1$ and $\mathbf{e}_2$ are linearly independent in a general case (e.g., for π polarization), that is, form a certain vector basis. It is clear that in the general case it is not possible to provide the field limitations in all the space if one has only the given system. Therefore, it is necessary to select the characteristic directions (or planes orthogonal to them) along which the field limitation is realized, that is, the obtained solution proves to be uniformly available. According to the physical meaning of the problem, one can assume that these directions are related to the transmitted ($\mathbf{\kappa}_{00}$) and diffracted ($\mathbf{\kappa}_h$) waves. This assumption is confirmed by the following.

The particular solution of the inhomogeneous equation

$$\mathrm{rotrot}\mathbf{E} - \kappa_0^2 \mathbf{E} = \mathbf{e}\exp\left(i\mathbf{\kappa}_0\mathbf{r}\right)$$

in the resonance case takes the form

$$\mathbf{E} = \left(\frac{i(\mathbf{\kappa}_0 \mathbf{r})}{2\kappa^2}\mathbf{e} - \frac{(\mathbf{\kappa}_0 \mathbf{e})}{\kappa^4}\mathbf{\kappa}_0 - \frac{i(\mathbf{\kappa}_0 \mathbf{e})(\mathbf{\kappa}_0 \mathbf{r})}{2\kappa^4}\mathbf{\kappa}_0\right)\exp\left(i\mathbf{\kappa}_0\mathbf{r}\right).$$

It is seen from here that the infinite increase in the wave amplitude is associated with the wave vector $\mathbf{\kappa}_0$ in the direction coinciding with $\mathbf{e}$. Then, according to the meaning of the zeroth approximation, it is necessary that the projections onto $\mathbf{e}_1$ in the first equation of Eq. (22) vanish and analogously the projections onto $\mathbf{e}_2$ in the second equation vanish.

Thus, multiplying scalarly the first equation of the system by $\mathbf{e}_1$ and the second one by $\mathbf{e}_2$, we obtain the scalar system

$$\begin{cases} i2(\nabla_1 c_1, \mathbf{\kappa}_{00}) + X c_1 + \kappa^2 \dfrac{\chi_{\overline{H}}}{\chi_H}\eta c_2 = 0 \\ \kappa^2 \eta c_1 + i2(\nabla_1 c_2, \mathbf{\kappa}_h) + X c_2 = 0 \end{cases} \tag{23}$$

where

$$\eta = (\mathbf{e}_1 \mathbf{e}_2) = \begin{cases} 1 - \text{for } \sigma \text{ polarization} \\ \cos 2\theta - \text{for } \pi \text{ polarization} \end{cases}$$

The obtained system (23) is virtually the dispersion relation written in the differential form for the transmitted and diffracted waves in the two-wave approximation. As follows from the presented conclusion, the possibility to obtain this system is dictated by the choice of the zeroth approximation. Namely, the wave vectors of transmitted and diffracted waves must have the identical component $\mathbf{\kappa}_{0\perp}$ that leads to the scattering interpretation as a result of reflection from the atomic plane.

We pass from the differential form of system (23) to the algebraic one; for this purpose we make the following substitution $c_j(\mathbf{r}_1) \rightarrow c_j \exp(i\mathbf{Pr})$, $\nabla_1 c_j = i\mathbf{P}c_j$, where $\mathbf{P}$ is a certain constant vector.

We obtain

$$
\begin{cases}
(-2(\mathbf{P}\boldsymbol{\kappa}_{00}) + X)c_1 + \kappa^2 \dfrac{\chi_{\overline{H}}}{\chi_H}\eta c_2 = 0 \\
\kappa^2 \eta c_1 + (-2(\mathbf{P}\boldsymbol{\kappa}_h) + X)c_2 = 0
\end{cases}
\tag{24}
$$

This system should be considered as the condition of field limitation in the specified direction of field propagation that is related to the structure parameters of the crystal and to the diffraction geometry. As for the physical meaning of the considered problem, this limitation should be in the direction of the normal inside crystal.

$$
\begin{cases}
(-2P\kappa_{00}\gamma_0 + X)c_1 + \kappa^2 \eta \dfrac{\chi_{\overline{H}}}{\chi_H} c_2 = 0 \\
\kappa^2 \eta c_1 + (-2P\kappa_h\gamma_h + X)c_2 = 0
\end{cases}
\tag{25}
$$

Here, $\gamma_{0,h}$ are the direction cosines of the corresponding wave vectors, $\kappa_h = \kappa_{00}$. The nontrivial solution of Eq. (25) requires the corresponding determinant to vanish:

$$
4\kappa_{00}^2 \gamma_n \gamma_0 P^2 - 2\kappa_{00} X(\gamma_0 + \gamma_h)P + X^2 - \kappa^4 \eta^2 \frac{\chi_{\overline{H}}}{\chi_H} = 0
\tag{26}
$$

The solution of quadratic equation relative to P takes the form

$$
P_{1,2} = \frac{X(\gamma_0 + \gamma_h) \pm \left[X^2(\gamma_0 - \gamma_h)^2 + 4\gamma_0\gamma_h\kappa^4\eta^2 \frac{\chi_{\overline{H}}}{\chi_H}\right]^{1/2}}{4\kappa_{00}\gamma_h\gamma_0} = \frac{X(\gamma_0 + \gamma_h) \pm D}{4\kappa_{00}\gamma_h\gamma_0}
\tag{27}
$$

Then, solving, for example, the first equation of system (25) relative to c_2, we obtain

$$
c_2 = \frac{X(\gamma_0 - \gamma_h) \pm D}{2\kappa^2 \eta \gamma_h} \cdot \frac{\chi_H}{\chi_{\overline{H}}} c_1 = \alpha_{1,2} c_1
\tag{28}
$$

Finally, going back to the initial variables, we represent the wave field in the crystal in the following form:

$$
\begin{aligned}
\mathbf{E} = {}& \exp{(i\chi_H(\mathbf{P_1 r}))}\left(\exp{(i\boldsymbol{\kappa}_{00}\mathbf{r})}\mathbf{e}_1 + \alpha_1 \exp{(i\boldsymbol{\kappa}_h\mathbf{r})}\mathbf{e}_2\right)c_{11} \\
& + \exp{(i\chi_H(\mathbf{P_2 r}))}\left(\exp{(i\boldsymbol{\kappa}_{00}\mathbf{r})}\mathbf{e}_1 + \alpha_2 \exp{(i\boldsymbol{\kappa}_h\mathbf{r})}\mathbf{e}_2\right)c_{12}
\end{aligned}
\tag{29}
$$

Here, the constants c_{ij} have the additional indices corresponding to the values of P_1 and P_2.

The equation for the wave field is simplified for the case of semi-infinite crystal when the wave reflection from a lower crystal face is eliminated. In this case the choice of sign in $P_{1,2}$ and correspondingly in $\alpha_{1,2}$ is defined by physical considerations. Namely, it is necessary that the wave field transfers to the usual form of the refracted wave propagating in the crystal as continuum at the deviation from the exact Bragg condition (increasing $|X|$). Then, we have

$$
P = \frac{X(\gamma_0 + \gamma_h) - \operatorname{sgn}{(X)}D}{4\kappa_{00}\gamma_h\gamma_0};
\tag{30}
$$

$$
\alpha = \frac{X(\gamma_0 - \gamma_h) - \operatorname{sgn}{(X)}D}{2\kappa^2 \eta \gamma_h \chi_{\overline{H}}/\chi_H}
\tag{31}
$$

$$\mathrm{sgn}\,(X) = \begin{cases} 1, & X \geq 0 \\ -1, & X < 0 \end{cases}$$

Finally, the wave field in the crystal takes the form

$$E = \exp\left(i\chi_H(\mathbf{Pr})\right)\left(\exp\left(i\kappa_{00}\mathbf{r}\right)\mathbf{e}_1 + \alpha \exp\left(i\kappa_h\mathbf{r}\right)\mathbf{e}_2\right)c \tag{32}$$

where the constant c is determined from the boundary conditions.

Eq. (32) describes the uniform wave field in a perfect crystal near the Bragg maximum.

To compare the obtained expression to the known results and to the experiment, it is necessary to go to the angular variable which connects with the deviation from the exact Bragg angle $\Delta\theta$. In this case, the parameter $X = 2(\kappa_{00}\kappa_{01})$ should be expressed by $\Delta\theta$. As was mentioned the parametric resonance (diffraction Laue condition $(\kappa_0\mathbf{h}) = \mp 1/2$) is determined only by the component $\kappa_{0\parallel}$ directed along $\mathbf{h}$, by the vector character of the problem. This means that the vector κ_{01} in expansion $\kappa_0 = \kappa_{00} + \chi_H\kappa_{01}$ is directed along $\mathbf{h}$, $\kappa_{01} = \kappa_{01}\mathbf{h}$. Taking into account that the reflection fixed in the experiment is determined by the normal component of the wave vector $\kappa_{0n} = (\kappa_0\mathbf{n})$, we obtain for X

$$X = -\frac{\kappa_{00}^2\beta_H\gamma_0}{(\gamma_h - \gamma_0)\chi_H} \tag{33}$$

Here, we introduce the standard (with an accuracy of refraction) angular variable:

$$\beta_H = -2\Delta\theta\sin 2\theta$$

As is known, two principal schemes are considered in the diffraction theory, by Laue ($\gamma_0 > 0$, $\gamma_h > 0$) and by Bragg ($\gamma_0 > 0$, $\gamma_h < 0$). In the case of Bragg diffraction, the wave field structure will qualitatively differ depending on the considered angular range.

In particular in the range

$$X^2(\gamma_0 + |\gamma_h|)^2 < 4\gamma_0|\gamma_h|\kappa^4\eta^2\frac{\chi_{\bar{H}}}{\chi_H},$$

the waves will exponentially attenuate along the normal surface into the crystal. In terms of the qualitative theory of differential equations, the solution will be unstable. The known interpretation [10, 11] leads to the conclusion of the expulsion of the wave from the crystal and the formation of a diffraction maximum. Thus, the indicated condition separates the stable solutions (oscillating type) from unstable ones (exponential type), that is, provides the equations of transition curves of the parametric plane (X, κ^2) [10, 11].

The width of the unstable region, the region of exponential wave attenuation, is determined by the expression

$$\Delta X = \frac{4\kappa^2\eta(\gamma_0|\gamma_h|)^{1/2}}{\gamma_0 + |\gamma_h|} \cdot \left(\frac{\chi_{\bar{H}}}{\chi_H}\right)^{1/2} \tag{34}$$

or proceeding to the angular variable

$$\Delta\theta = \frac{2\eta\left(\chi_H\chi_{\bar{H}}\right)^{1/2}}{(1+\chi_0)\sin 2\theta} \cdot \left(\frac{|\gamma_h|}{\gamma_0}\right)^{1/2} \tag{35}$$

This is the known expression (with an accuracy of refraction) for the angular width of the Bragg table for the case of the semi-infinite perfect non-absorbing crystal. The extinction length Λ_{ext} is determined as a decrement of wave attenuation in the point Bragg position:

$$\Lambda_{ext} = \frac{2\kappa_{00}\left(|\gamma_h|\gamma_0\right)^{1/2}}{\kappa^2\eta\left(\chi_{\bar{H}}\chi_H\right)^{1/2}} \approx \frac{\kappa}{|\chi_H|} \tag{36}$$

We now summarize the intermediate stage. The use of the generalized method of many scales allowed us to obtain the system of basic equations describing the behavior of wave field near the Bragg maximum in the two-wave approximation. This system is a direct analog of the dispersion relations of the Ewald-Laue theory and the Takagi-Taupin system of the generalized dynamic theory. The substantial difference of the developed variant of the theory consists in the cancelation of the shortening procedure of equations by neglecting the second derivatives. The principal moment here is the expansion in χ_H which makes it possible to save maximally the structure of Maxwell equations for the wave field in the crystal under conditions of the dynamic diffraction.

Comparison of the obtained results to that known from the Takagi-Taupin theory shows the complete correspondence both in the qualitative interpretation of types of the solutions obtained in the different angular ranges in the case of Bragg diffraction and in the analytical expressions for the width of the Bragg maximum and the extinction length. This correspondence indicates that in spite of the formal representation of $\chi(\mathbf{r})$ in the form of infinite Fourier series in the Ewald-Laue and Takagi-Taupin theories only three terms of the series are really used.

However, the value of the theory developed here shows itself to a large degree when the boundary conditions are taken into account that takes the explicit expressions for the reflection coefficient. Therefore, we consider now the boundary conditions and determine the principal difference between our approach and known variants of the dynamical diffraction theory.

2.4 Boundary conditions and the amplitude reflection coefficient

According to Eq. (32), the expression above obtained for the wave field in the crystal depends on the constant c, which must be determined by the boundary conditions of the problem. As was indicated above, it is not possible to use the classical boundary conditions of electrodynamics in the Takagi-Taupin theory since negligibility of the second derivatives of amplitudes with respect to the coordinates reduces the order of the equation. As a result the boundary conditions redefine the problem, and the new boundary conditions are stated that determine only the field amplitudes on a crystal surface. This procedure proves to be quite correct for usual diffraction geometries, when the angles of incidence and yield of waves substantially exceed the critical values.

We find out now the differences appearing when the boundary conditions are strictly taken into account in our theory.

The reflection coefficient is determined as a ratio of the averaged values of normal components of the pointing vector for the diffracted and incident waves:

$$R = \left|\frac{c_h^0}{c^0}\right|^2 \cdot \frac{\left(\mathbf{n}\boldsymbol{\kappa}_h^0\right)}{\left(\mathbf{n}\boldsymbol{\kappa}\right)} \tag{37}$$

Here, $\mathbf{n}$ is the unit vector of the normal directed inside the crystal, and $\boldsymbol{\kappa}$, c^0 and $\boldsymbol{\kappa}_h^0$, c_h^0 are the wave vectors and amplitudes of the incident and diffracted waves, respectively. Index 0 denotes that the values of the indicated quantities are related to the environment vacuum. Thus, the determination of the reflection coefficient is associated with the determination of the diffracted wave amplitude in the vacuum. This problem is solved by the boundary conditions.

As is known, the boundary conditions require continuity of the tangential components of electric and magnetic fields which is a consequence of the uniformity of the problem along the surface. The boundary problem disintegrates into successive steps related to the determination of c_h^0. In this case the elementary problem is considered at each stage, namely, the determination of the relation between the amplitudes of incident, transmitted, and secularly reflected waves. The solution of this problem leads to the known Fresnel formulas:

$$c = \frac{(\mathbf{n}, \mathbf{\kappa}) - (\mathbf{n}, \mathbf{\kappa}_R)}{(\mathbf{n}, \mathbf{\kappa}_0 + \mathbf{P}) - (\mathbf{n}, \mathbf{\kappa}_R)} c^0 = \frac{2(\mathbf{n}, \mathbf{\kappa})}{(\mathbf{n}, \mathbf{\kappa}_0 + \mathbf{P}) + (\mathbf{n}, \mathbf{\kappa})} c^0 \tag{38}$$

$$c_R = \frac{(\mathbf{n}\mathbf{\kappa}) - (\mathbf{n}, \mathbf{\kappa}_0 + \chi_H \mathbf{P})}{(\mathbf{n}, \mathbf{\kappa}_0 + \chi_H \mathbf{P}) - (\mathbf{n}\mathbf{\kappa}_R)} c^0 = \frac{(\mathbf{n}\mathbf{\kappa}) - (\mathbf{n}, \mathbf{\kappa}_0 + \chi_H \mathbf{P})}{(\mathbf{n}, \mathbf{\kappa}_0 + \chi_H \mathbf{P}) + (\mathbf{n}\mathbf{\kappa})} c^0 \tag{39}$$

$$c_h^0 = \frac{(\mathbf{n}, \mathbf{\kappa}_h + \chi_H \mathbf{P}) - (\mathbf{n}, \mathbf{\kappa}_{hR})}{(\mathbf{n}, \mathbf{\kappa}_h^0) - (\mathbf{n}, \mathbf{\kappa}_{hR})} \alpha c = \frac{(\mathbf{n}, \mathbf{\kappa}_h + \chi_H \mathbf{P}) - (\mathbf{n}, \mathbf{\kappa}_{hR})}{(\mathbf{n}, \mathbf{\kappa}_h^0) - (\mathbf{n}, \mathbf{\kappa}_{hR})} \cdot \frac{2(\mathbf{n}, \mathbf{\kappa})}{(\mathbf{n}, \mathbf{\kappa}_0 + \chi_H \mathbf{P}) + (\mathbf{n}, \mathbf{\kappa})} \alpha c^0 \tag{40}$$

Here, c_R is the amplitude of the specularly reflected wave, and $\mathbf{\kappa}_{hR}$ is the wave vector of the diffraction wave specularly reflected from the lower side of the crystal-vacuum interface.

The obtained relations allow us to determine not only the diffracted wave but also the specularly reflected wave, which principally distinguishes our approach from the formalism of the Takagi-Taupin equations.

Eqs. (38)–(40) solve the problem of the determination of field amplitudes under the conditions of sliding noncoplanar diffraction when the incident and diffracted waves are near the critical angles of total external reflection (TER). They are analogous to the relations obtained in [12] where this problem was solved by the fourth-order dispersion equation.

The correspondence with the Takagi-Taupin theory must be undoubtedly fulfilled for the case of large angles of incidence and yield of the diffraction wave. Really in this case, the amplitude of specular wave tends to be zero, and the reflection coefficient takes the form

$$R = \left| \frac{c_h^0}{c^0} \right|^2 \cdot \frac{(\mathbf{n}\mathbf{\kappa}_h^0)}{(\mathbf{n}\mathbf{\kappa})} = |\alpha|^2 \frac{(\mathbf{n}\mathbf{\kappa}_h^0)}{(\mathbf{n}\mathbf{\kappa})}$$

$$= \frac{\gamma_0(1 + \chi_0)\beta_H - \text{sgn}(\beta_H) \left[(\gamma_0(1 + \chi_0)\beta_H)^2 + 4\gamma_0\gamma_H \eta^2 \chi_{\overline{H}}\chi_H \right]^{1/2}}{2\gamma_H \eta \chi_H} \cdot \frac{\gamma_H}{\gamma_0} \tag{41}$$

This is the known expression for the coefficient of reflection from a perfect half-infinite crystal, i.e. the Bragg table. In addition to this in the case of extremely asymmetric diffraction when the diffraction wave leaving the crystal is almost parallel to the surface, the amplitude is modulated by the factors taking into account the refraction of transmitted and diffracted waves at the crystal-vacuum interface and the diffraction wave interaction related to the vector $\mathbf{P}$.

The proposed covariant (it may be named nonstandard) theory allows the generalization for the case of the crystal with lattice deformations. Therefore, a uniform approach to the account of deformations and other distortions in all diffraction schemes is realized.

3. Generalization of a covariant dynamic diffraction theory to the case of deformed crystal

It is known that Takagi-Taupin equations were obtained using the model concept of the character of lattice distortions which makes it possible to directly take into account the displacement of atomic planes in the three-dimensional periodic function of crystal polarizability. This concept allows for describing lattice displacements and strain using the methods of classical theory of elasticity formulated within the continuum approximation. To satisfy the condition of the dynamic character of scattering in the Takagi-Taupin theory, the lattice distortion is assumed to be rather weak; correspondingly, the strain is small. The character of variation in strain is implicitly taken into account only when the wave field is chosen in the form of a Bloch function with slowly varying amplitudes; thus, the question of applicability of this concept remains open.

There is another limitation of the Takagi-Taupin theory which is related to the correct statement of boundary conditions. Mathematically, the Takagi-Taupin equations form a first-order differential system with respect to the scalar amplitudes of transmitted and diffracted waves. The procedure of determining these amplitudes at the crystal-vacuum interface does not correspond to the classical boundary conditions.

This discrepancy is due to the fact that the Takagi-Taupin equations are obtained disregarding the second derivatives of the field amplitudes with respect to coordinates. Thus, the boundary conditions impose fundamental limitations on the applicability of the Takagi-Taupin equations (e.g., when analyzing extremely asymmetric diffraction schemes).

In this section we generalized the covariant theory of dynamic diffraction which was presented in the previous section, to a crystal with a distorted lattice. This approach makes it possible to formulate the limitation on the character of variation in strain for the applicability of the Takagi-Taupin equations. In addition, the equations obtained can be applied (as in the case of an ideal crystal) for arbitrary diffraction schemes.

In this section, we follow the original papers [13, 14].

3.1 Wave field in the absence of diffraction

The polarizability $\chi(\mathbf{r'})$ of a crystal with a distorted lattice is a function of coordinates; however, in contrast with an ideal crystal, it depends not only on the reciprocal lattice vector $\mathbf{H}$ but also on the vector of atomic plane displacement from equilibrium $\mathbf{u}(\mathbf{r'})$. According to the accepted assumptions of the generalized dynamic theory, we choose the crystal model $\chi(\mathbf{r'})$ in the form

$$\chi(\mathbf{r'}) = \chi_0 + \chi_H \exp\left(i\mathbf{H}(\mathbf{r'} + \mathbf{u}(\mathbf{r'}))\right) + \chi_{\overline{H}} \exp\left(-i\mathbf{H}(\mathbf{r'} + \mathbf{u}(\mathbf{r'}))\right) \tag{42}$$

It can be seen that the general structure of $\chi(\mathbf{r'})$ corresponds to the case of an ideal crystal. Obviously, this situation is possible only when the displacement $\mathbf{u}(\mathbf{r'})$ is small. Then, for Eq. (7) we arrive at

$$\text{rotrot}\mathbf{E}(\mathbf{r'}) - k^2\left(1 + \chi_0 + \chi_H \exp\left(i\mathbf{H}(\mathbf{r'} + \mathbf{u}(\mathbf{r'}))\right) + \chi_{\overline{H}} \exp\left(-i\mathbf{H}(\mathbf{r'} + \mathbf{u}(\mathbf{r'}))\right)\right)\mathbf{E}(\mathbf{r'}) = 0 \tag{43}$$

As in the case of an ideal crystal, Eq. (43) should be reduced to a dimensionless form. To this end we will use the length $\mathbf{H}$ of the reciprocal lattice vector:

$$\text{rotrot}\mathbf{E}(\mathbf{r}) - \kappa^2\left(1 + \chi_0 + \chi_H \exp\left(i\mathbf{h}(\mathbf{r} + \mathbf{u}(\mathbf{r}))\right)\right) + \chi_{\overline{H}} \exp\left(-i\mathbf{h}(\mathbf{r} + \mathbf{u}(\mathbf{r}))\right)\mathbf{E}(\mathbf{r}) = 0;$$

$$(44)$$

The approximate solution to Eq. (44) in the Takagi-Taupin theory is known to be sought after in the form of plane waves with slowly varying amplitudes. Finally, the Takagi-Taupin equations can be derived from Eq. (44) with allowance for the two-wave approximation [2].

The covariant dynamic theory in the case of an ideal crystal is due to the fact that the Fourier component of polarizability χ_H which is responsible for the excitation of a diffraction wave in the crystal under the corresponding Laue geometric condition for the wave vectors of refracted and diffracted waves and the reciprocal lattice vector was chosen to be the perturbation parameter. The parameter χ_0 cannot be considered a perturbation parameter because it leads to only refraction of the incident wave in the crystal and does not influence the occurrence of diffraction effects. Obviously, in the case of a weakly deformed crystal, the criterion of the choice of χ_H as the expansion parameter remains valid because the presence of a weak displacement field only transforms the diffraction pattern rather than breaking it. This circumstance allows one to extend (with necessary modifications) the scheme of constructing a solution in the nonstandard approach to a deformed structure.

Recall that in the case of an ideal crystal the direct expansion of the solution to Eq. (44) in the parameter χ_H is inconsistent when a parametric resonance is observed which corresponds to the Laue diffraction condition:

$$(\kappa_0 \pm \mathbf{h})^2 - \kappa_0^2 = 0 \qquad (45)$$

In this case, the scattered wave amplitude increases unlimitedly. This is due to the fact that the diffraction wave can be found as a particular solution to the inhomogeneous wave equation:

$$\text{rotrot}\mathbf{E} - \kappa^0{}_2\mathbf{E} = \mathbf{E}_0 \exp\left(i(\kappa_0 \pm \mathbf{h})\mathbf{r}\right) \qquad (46)$$

which according to Eq. (13) has the form

$$\mathbf{E} = \frac{\mathbf{E}_0 \exp\left(i(\kappa_0 \pm \mathbf{h})\mathbf{r}\right)}{(\kappa_0 \pm \mathbf{h})^2 - \kappa_0^2} \qquad (47)$$

For a deformed crystal, the plane wave is replaced with $\mathbf{E}_0\exp(i(\mathbf{qr} \pm \mathbf{hu}(\mathbf{r})))$ within model (42); as a result a particular solution to Eq. (46) cannot be written in the form as in Eq. (47). A particular solution (Eq. (47)) can generally be represented in the integral form using Green's function for the corresponding homogeneous equation. However, the generality of this representation is devalued by the difficulties in analyzing the relations (e.g., in view of the vector character of the problem Green's function has generally speaking a tensor form). As a result the integral representation of the solution to Eq. (46) is basically formal.

At the same time from the physical point of view, the displacement field changes significantly at distances much larger than the lattice parameter. This limitation is substantiated in particular by the fact that we describe lattice distortions within the continuum approximation. In this case the approximate solution to Eq. (46) can be obtained similarly to Eq. (47):

$$E = \frac{E_0 \exp\left(i(\kappa_0 \pm h)r \pm ihu(r)\right)}{(\kappa_0 \pm h)^2 - \kappa_0^2} \tag{48}$$

Correspondingly, the direct expansion of the solution to Eq. (44) for a deformed crystal is similar to that for an ideal crystal and is determined by the wave superposition in the form

$$E_{01(2)}\left(\exp\left(i\kappa_0 r\right) + \kappa^2 \chi_{H(\overline{H})} \frac{\exp\left(i(\kappa_0 \pm h)r \pm ihu(r)\right)}{(\kappa_0 \pm h)^2 - \kappa_0^2}\right) \tag{49}$$

where E_{01} and E_{02} are the amplitudes of the plane waves $\exp.(\pm i\kappa_0 r)$ propagating in the crystal, considering a continuous medium with $\chi = 1 + \chi_0$. This solution describes the wave field in the crystal with a distorted lattice beyond the angular ranges of diffraction reflection (i.e., in the nonresonant case). The question of the accuracy of approximation (49) remains open until the character of the change in $u(r)$ is specified.

The most general limitation on the strain in the crystal is imposed by the requirement for the smallness of the strain tensor ε corresponding to a given displacement field:

$$\varepsilon = \varepsilon_0 \frac{\partial u(r)}{\partial r} \ll 1, \tag{50}$$

where ε_0 is the strain amplitude in the structure. For sufficiently regular displacement fields, this requirement is reduced to the condition $\varepsilon_0 \ll 1$. It follows from Eq. (50) that in the case of a deformed crystal when the aforementioned conditions are satisfied the applicability of direct expansion remains limited because of the Laue resonant condition (45) where the amplitudes of waves excited in the crystal increase unlimitedly. Thus, the method for obtaining an approximate solution must also be modified in the case of a crystal with a distorted lattice. The choice of the modification technique depends on the character of the displacement field in the crystal and obviously cannot provide a universal solution for all physically possible cases. We will consider the most widespread situation where the displacement field changes at distances comparable with the extinction length. In this case the multiscale method which is the basis of the covariant theory of diffraction in an ideal crystal can directly be extended to a deformed structure.

3.2 Derivation of the main equations for strained crystal

The strain field in a crystal may have various forms depending on the nature of lattice distortions. These forms can mathematically be represented by setting different structural parameters (e.g., the thicknesses of epitaxial layers and transition regions between layers, sizes of lattice-strain regions caused by various defects, etc.). The values of these parameters are determined by not only the strain amplitudes ε_{0i} but also the characteristic regions L_i of their variation in the crystal. As a result this situation can symbolically be presented in a form that explicitly relates the parameters:

$$Hu = F\left(\frac{\varepsilon_{0i} L_i}{d}, \frac{rd}{L_i}\right) \tag{51}$$

Here, $\varepsilon_{0i}L_i/d$ is the amplitude of the displacement field, and rd/L_i is the size of the distorted region in the accepted coordinate normalization (d is the interplanar spacing). The influence of the factor $\mathbf{Hu}$ on the diffraction effects should correlate with χ_H which physically determines the characteristic region of wave-field formation under dynamic diffraction conditions. This concerns both the angular range of diffraction reflection and the field extinction length in the crystal.

Thus, the parameter χ_H must implicitly be taken into account in functional relation (51) which takes the form

$$\mathbf{Hu} = F\left(\frac{\varepsilon_0}{m_i \chi_H^i}, m_i \mathbf{r}_i\right), \quad m_i = \frac{1}{HL\chi_H^i} = \frac{d}{2\pi L \chi_H^i}, \mathbf{r}_i = \chi_H^i \mathbf{r}. \tag{52}$$

The powers i in Eq. (52) can be integers; however, fractional values (e.g., 1/2) are physically most interesting because they indicate changes in displacement fields on scales below the extinction length.

Thus, a consideration of different types of lattice distortions generally calls for taking into account different characteristic spatial regions; this approach completely corresponds to the main concept of the multiscale method [9]. Obviously, different modifications of the method are required depending on the specific structure of Eq. (52). We will consider the simplest case of one scale which leads in a particular case to Takagi-Taupin equations.

Let us consider the atomic plane displacement occurring at some effective layer thickness L. The parameter $\mathbf{Hu}(\mathbf{r})$ can be written as

$$\mathbf{Hu}(\mathbf{r}) = \varepsilon_0 LHF\left(\frac{\mathbf{r}}{L}\right), \tag{53}$$

where $F\left(\frac{\mathbf{r}}{L}\right)$ is the displacement model and ε_0 is the strain amplitude.

According to the multiscale method [9], the main equation (Eq. (44)) is analyzed near the Bragg maximum on different spatial scales (determined by χ_H) using the transition from one variable $\mathbf{r}$ to several variables $\mathbf{r}_0 = \mathbf{r}$, $\mathbf{r}_1 = \chi_H \mathbf{r}$, $\mathbf{r}_2 = \chi_H^2 \mathbf{r}$, ... If one seeks the first-order approximation for χ_H, two scales (affecting the field in the crystal) should be considered:

$$E(\mathbf{r}) = E(\mathbf{r}_0, \mathbf{r}_1).$$

Then, according to Section 2, the field, the operator rot, and the wave vector κ_0 are expanded in χ_H powers:

$$E = E_0(\mathbf{r}_0, \mathbf{r}_1) + \chi_H E_1(\mathbf{r}_0, \mathbf{r}_1) + ...;$$

$$\kappa_0 = \kappa_{00} + \chi_H \kappa_{01} + ...;$$

$$\kappa_0^2 = (\kappa_0, \kappa_0) = \kappa_0^2 + 2\chi_H(\kappa_0, \kappa_{01}) + ... = \kappa_0^2 + \chi_H X; \quad X = 2(\kappa_0, \kappa_{01})$$

$$\mathrm{rotrot} = \mathrm{rot}_0\mathrm{rot}_0 + \chi_H(\mathrm{rot}_0\mathrm{rot}_1 + \mathrm{rot}_1\mathrm{rot}_0) + ...$$

(54)

Substituting expansion (54) into the main equation (Eq. (44)), we obtain

$$(\mathrm{rot}_0\mathrm{rot}_0 + \chi_H\mathrm{rot}_1\mathrm{rot}_0 + \chi_H\mathrm{rot}_0\mathrm{rot}_1 + ...)(E_0 + \chi_H E_1 + ...) -$$
$$-(\kappa_{00}^2 + \chi_H X + ... + \kappa^2\chi_H(\exp(i\mathbf{h}\mathbf{r}_0 + i\phi(\mathbf{r}_1)) +$$
$$\frac{\chi_{\overline{H}}}{\chi_H}\exp(-i\mathbf{h}\mathbf{r}_0 - i\phi(\mathbf{r}_1))))(E_0 + \chi_H E_1 + ...) = 0$$

(55)

In contrast to the case of an ideal crystal, $\chi(\mathbf{r})$ is now presented as a function of two spatial scales: $\mathbf{r}_0$ and $\mathbf{r}_1$. Here, the following designation is introduced:

$$\phi(\mathbf{r}_1) = \frac{\varepsilon_0}{m\chi_H}\mathbf{Hu}(m\mathbf{r}_1), \quad m = \frac{1}{HL\chi_H} = \frac{d}{2\pi L\chi_H}. \tag{56}$$

The parameter $1/m$ means some effective thickness of the deformed layer on the scale $\mathbf{r}_1$. The possibility of presenting the displacement as a function of the scale $\mathbf{r}_1$ suggests the condition $m \sim 1$, i.e., the number d of interplanar spacings embedded in the deformed layer thickness should be on the order of the extinction length Λ_{ext} on the dimensionless scale $\mathbf{r}$:

$$\frac{L}{d} \sim \frac{1}{\chi_H} \sim \Lambda_{ext} \tag{57}$$

We will follow the scheme for solving Eq. (55) that was reported in Section 1. As for an ideal crystal, the initial approximation can be written in the form as in Eq. (18). Accordingly, the operator rot acts on only one spatial scale $\mathbf{r}_0$. Since we are interested in the wave field near the Bragg maximum, a solution to Eq. (18) within the two-wave approximation should be sought after in the form of a superposition of transmitted and diffracted waves:

$$\mathbf{E}_0 = \mathbf{e}_1 c_1(\mathbf{r}_1) \exp(i\boldsymbol{\kappa}_{00}\mathbf{r}_0) + \mathbf{e}_2 c_2(\mathbf{r}_1) \exp(i\boldsymbol{\kappa}_h \mathbf{r}_0)$$

$$\boldsymbol{\kappa}_h = \boldsymbol{\kappa}_0 + \mathbf{h}, \tag{58}$$

The first-order approximation with respect to χ_H yields the inhomogeneous equation:

$$\text{rot}_0\text{rot}_0\mathbf{E}_1 - \kappa_{00}^2\mathbf{E}_1 = -\,(\text{rot}_1\text{rot}_0 + \text{rot}_0\text{rot}_1)\mathbf{E}_0 + X\mathbf{E}_0$$

$$+ \kappa^2\left(\exp(i\mathbf{hr}_0 + i\phi(\mathbf{r}_1)) + \frac{\chi_{\bar{H}}}{\chi_H}\exp(-i\mathbf{hr}_0 - i\phi(\mathbf{r}_1))\right)\mathbf{E}_0 \tag{59}$$

Hence

$$\text{rot}_0\text{rot}_0\mathbf{E}_1 - \kappa_{00}^2\mathbf{E}_1 = \left((i2(\nabla_1 c_1, \boldsymbol{\kappa}_{00}) + Xc_1)\mathbf{e}_1 + \kappa^2\frac{\chi_{\bar{H}}}{\chi_H}c_2\mathbf{e}_2\exp(-i\phi(\mathbf{r}_1))\right)\exp(i\boldsymbol{\kappa}_{00}\mathbf{r}_0) +$$

$$+\;(\kappa^2 c_1\mathbf{e}_1\exp(i\phi(\mathbf{r}_1)) + (i2(\nabla_1 c_2, \boldsymbol{\kappa}_h) + Xc_2)\mathbf{e}_2)\exp(i\boldsymbol{\kappa}_h\mathbf{r}_0) +$$

$$+\;\kappa^2\frac{\chi_{\bar{H}}}{\chi_H}\exp(i(\boldsymbol{\kappa}_{00} - \mathbf{h})\mathbf{r}_0)\exp(-i\phi(\mathbf{r}_1))c_1\mathbf{e}_1 + \kappa^2\exp(i(\boldsymbol{\kappa}_h + \mathbf{h})\mathbf{r}_0)\exp(i\phi(\mathbf{r}_1))c_2\mathbf{e}_2.$$

$$\tag{60}$$

Here, the gradient ∇_1 acts in the space $\mathbf{r}_1$. Eq. (60) was derived taking into account the additional condition $(\nabla_1 c_1, \mathbf{e}_1) = 0$. This indicates that the change in amplitudes on the scale $\mathbf{r}_1$ does not violate the wave transversity.

As can easily be found from Eq. (47), two first terms in the right-hand part of Eq. (60) generate the divergence of a particular solution. The following vector system can be obtained by excluding these terms from Eq. (60):

$$\begin{cases} (i2(\nabla_1 c_1, \boldsymbol{\kappa}_{00}) + Xc_1)\mathbf{e}_1 + \kappa^2\frac{\chi_{\bar{H}}}{\chi_H}\exp(-i\phi(\mathbf{r}_1))c_2\mathbf{e}_2 = 0 \\[2mm] (i2(\nabla_1 c_2, \boldsymbol{\kappa}_h) + Xc_2)\mathbf{e}_2 + \kappa^2\exp(i\phi(\mathbf{r}_1))c_1\mathbf{e}_1 = 0 \end{cases} \tag{61}$$

According to Section 2, to satisfy the field boundedness condition, the projections of Eq. (61) on the corresponding unit vectors $\mathbf{e}_1$ and $\mathbf{e}_2$ must be nullified. Then, we obtain the following scalar system from Eq. (61):

$$\begin{cases} i2(\nabla_1 c_1, \mathbf{\kappa}_{00}) + Xc_1 + \kappa^2 \dfrac{\chi_{\overline{H}}}{\chi_H} \eta \exp\left(-i\phi(\mathbf{r}_1)\right)c_2 = 0 \\ i2(\nabla_1 c_2, \mathbf{\kappa}_h) + \kappa^2 \eta \exp\left(i\phi(\mathbf{r}_1)\right)c_1 + Xc_2 = 0 \end{cases} \tag{62}$$

According to Section 2, the parameter $X = 2(\mathbf{\kappa}_{00}\mathbf{\kappa}_{01})$ can be expressed in terms of the deviation from the exact Bragg angle $\Delta\theta$ as follows:

$$X = \frac{\gamma_0 \kappa_{00}^2 \beta_H}{(\gamma_h - \gamma_0)\chi_H} \tag{63}$$

Eq. (62) describes the changes in the wave amplitudes on the scale $\mathbf{r}_1$ in the Bragg reflection range for a crystal with a specified displacement field $\mathbf{u}(\mathbf{r}_1)$. Thus, the amplitudes are slowly varying parameters. This condition is in fact the basis of the formalism of generalized dynamic theory which results in the Takagi-Taupin equations. In this case the system of Eq. (62) should be considered a direct analog of the Takagi-Taupin equations. Let us prove this statement. We make the following substitutions in Eq. (62):

$$c_1 \rightarrow c_1 \exp\left(i a \mathbf{r}_1\right), \quad c_2 \rightarrow c_2 \exp\left(i a \mathbf{r}_1\right)$$

where $\mathbf{a} = a\mathbf{n}$ is a constant vector directed along the normal to the crystal surface. Vector $\mathbf{a}$ is chosen so as to make parameter X absent in the first equation of system (62). Then, a can be written as

$$a = \frac{X}{2\kappa_{00}\gamma_0}$$

Accordingly, Eq. (62) is reduced to the form

$$\begin{cases} i2(\nabla_1 c_1, \mathbf{\kappa}_{00}) + \kappa^2 \dfrac{\chi_{\overline{H}}}{\chi_H} \eta \exp\left(-i\phi(\mathbf{r}_1)\right)c_2 = 0 \\ i2(\nabla_1 c_2, \mathbf{\kappa}_h) + \kappa^2 \eta \exp\left(i\phi(\mathbf{r}_1)\right)c_1 - \dfrac{\kappa_{00}^2 \beta_H}{\chi_H}c_2 = 0 \end{cases} \tag{64}$$

In a particular case of coplanar diffraction in the xz plane oriented normally to the crystal surface, the gradient in Eq. (64) can be written as

$$\nabla_1 = \frac{\partial}{\partial x_1} + \frac{\partial}{\partial z_1}$$

Under the assumption that the displacement field in Eq. (64) depends on only x and z coordinates in the plane of incidence, system (64) is transformed into the set of Takagi-Taupin equations written on the scale $\mathbf{r}_1 = \chi_H \mathbf{r}$. Thus, the formalism considered here which is based on applying the multiscale method is in complete agreement with the generalized Takagi-Taupin dynamical theory of diffraction [2]. However, there is an important difference. The reduction of the covariant theory for a deformed crystal to the Takagi-Taupin equations in a particular case suggests that key condition (57) is satisfied, i.e., this correspondence is valid for only displacement fields changing in a region comparable in size with the extinction length.

Obviously, other displacement fields which a fortiori do not satisfy condition (57) can be implemented in real crystals. In this case the use of Takagi-Taupin equations may be unjustified and in any case should be additionally analyzed.

3.3 Reflection coefficient

System (64) has a more general character because it was obtained without additional limitations on the Takagi-Taupin equations which is related to the rejection of the second derivatives of the field amplitudes with respect to coordinates. It is known that the boundary conditions at the crystal-vacuum interface cannot be correctly taken into account due to these limitations. Finally, the Takagi-Taupin equations cannot be applied to extremely asymmetric diffraction schemes.

Let us show how the consideration of the boundary conditions yields explicit expressions for the amplitudes of diffracted and specularly reflected waves for arbitrary angles of incidence. For simplicity we will consider diffraction in a semi-infinite crystal in the case of σ polarization where the vectors $\mathbf{e}_1$ and $\mathbf{e}_2$ are mutually parallel. Here, the solution providing extinction wave decay near the Bragg maximum should be chosen from two linearly independent solutions to Eq. (62). Correspondingly, the solution to Eq. (62) for $c_1(\mathbf{r}_1)$ by analogy with an ideal crystal (Section 2) can be written as

$$c_1(\mathbf{r}_1) = c \exp{(i\mathbf{P}(\mathbf{r}_1)\mathbf{r}_1)} \tag{65}$$

where the constant c can be found from the boundary conditions and the vector $\mathbf{P}$, in contrast to Section 2, is assumed to be variable on the scale $\mathbf{r}_1$. Then, the amplitude $c_1(\mathbf{r}_1)$ according to the first equation of system (62) is determined as

$$c_2(\mathbf{r}_1) = \frac{\chi_H}{\kappa^2 \chi_{\bar{H}}} (2(\nabla_1(\mathbf{P}\mathbf{r}_1)\kappa_{00}) - X) \exp{(i\mathbf{P}\mathbf{r}_1)} \exp{(i\phi(\mathbf{r}_1))}c = \alpha c \tag{66}$$

Finally, the wave field in the semi-infinite crystal near the Bragg angle of incidence can be described by an expression that formally corresponds to an ideal crystal:

$$\mathbf{E} = \exp{(i\mathbf{P}\mathbf{r}_1)}(\exp{(i\boldsymbol{\kappa}_{00}\mathbf{r})} + \alpha \exp{(i\boldsymbol{\kappa}_h\mathbf{r})})c\mathbf{e} \tag{67}$$

The procedure of solving the boundary problem which can be reduced to a successive establishment of relations between wave amplitudes having common tangential components of the electric and magnetic fields at the crystal-vacuum interface remains the same. Therefore, one can use the expressions for amplitudes obtained in Section 2. As a result we arrive at formulas that are similar to the Fresnel formulas:

$$c = \frac{2(\mathbf{n}, \boldsymbol{\kappa})}{(\mathbf{n}, \boldsymbol{\kappa}_0 + \chi_H \mathbf{P}(\chi_H \mathbf{r}_s)) + (\mathbf{n}, \boldsymbol{\kappa})} c^0 \tag{68}$$

$$c_R = \frac{(\mathbf{n}\boldsymbol{\kappa}) - (\mathbf{n}, \boldsymbol{\kappa}_0 + \chi_H \mathbf{P}(\chi_H \mathbf{r}_s))}{(\mathbf{n}, \boldsymbol{\kappa}_0 + \chi_H \mathbf{P}(\chi_H \mathbf{r}_s)) + (\mathbf{n}\boldsymbol{\kappa})} c^0 \tag{69}$$

$$c_h^0 = \frac{(\mathbf{n}, \boldsymbol{\kappa}_h + \chi_H \mathbf{P}(\chi_H \mathbf{r}_s)) - (\mathbf{n}, \boldsymbol{\kappa}_{hR})}{(\mathbf{n}, \boldsymbol{\kappa}_h^0) - (\mathbf{n}, \boldsymbol{\kappa}_{hR})} \cdot \frac{2(\mathbf{n}, \boldsymbol{\kappa})}{(\mathbf{n}, \boldsymbol{\kappa}_0 + \chi_H \mathbf{P}(\chi_H \mathbf{r}_s)) + (\mathbf{n}, \boldsymbol{\kappa})} \alpha c^0 \tag{70}$$

The following designations are introduced here: $\mathbf{n}$ is the unit vector directed along the normal to the crystal surface into the crystal bulk; $\boldsymbol{\kappa}(c^0)$ and $\boldsymbol{\kappa}_h^0 (c_h^0)$ are the wave vectors (amplitudes) of the incident and diffracted waves, respectively; c_R is the amplitude of the specularly reflected wave; and $\mathbf{\kappa}_{hR}$ is the wave vector of the diffraction wave which is specularly reflected from the lower side of the crystal-vacuum interface. The superscript "0" is used for the parameters related to the environment. Vector $\mathbf{r}_s$ lies in the plane of crystal-vacuum interface. In Eqs. (68)–(70), we returned to the initial dimensionless variable $\mathbf{r}$.

Eqs. (68)–(70) can be used to find the field amplitudes for arbitrary angles of incidence including those in the vicinity of the critical total reflection angles. In particular the formula for reflectance can be found from Eq. (70) as follows:

$$R = \left|\frac{c_h^0}{c^0}\right|^2 \cdot \frac{(\mathbf{n}\mathbf{\kappa}_h^0)}{(\mathbf{n}\mathbf{\kappa})} \tag{71}$$

In a particular case of large angles of incidence, Eq. (71) can be simplified, and Eq. (70) yields the following expression for the reflectance (which can be derived from the Takagi-Taupin equations):

$$R = \left|\frac{c_h^0}{c^0}\right|^2 \cdot \frac{(\mathbf{n}\mathbf{\kappa}_h^0)}{(\mathbf{n}\mathbf{\kappa})} = |\alpha|^2 \frac{\gamma_H}{\gamma_0} \tag{72}$$

4. Conclusions

The variant of the dynamic X-ray diffraction presented in the present work is based on direct analysis of Maxwell equations for the definite model representations of the field-medium interaction taking into account the lattice presence which agree as a whole with the Ewald-Laue theory. This analysis proves to be available when the method of many scales adapted to the vector character of the problem is used. In this case the magnitude χ_H is the parameter of expansion that corresponds in full to the physical character of the problem. This correspondence is reflected in the mathematical structure of the analyzed field equation in the crystal under conditions of dynamic scattering.

The expressions obtained for the main field characteristics in the Bragg maximum region following from the qualitative singularities of the field propagation correspond to the known results of the dynamical theory. However, the correct use of the boundary conditions leads to an expression for the reflection coefficient that substantially differs from the classical one for the case of extremely asymmetric diffraction schemes. In addition the presented approach provides the amplitude of specularly reflected wave under conditions of dynamic diffraction, which cannot be apparently obtained in the framework of traditional approaches.

In the present work, we do not state the problem to analyze the features of dynamic scattering in the sliding diffraction geometry.

In conclusion, we note the most important in our opinion differences and advantages of the approach developed in the present work.

The second-order wave equation analyzed without any additional assumptions of the possibility of the interaction of refracted and scattered waves automatically results in dynamical scattering character; in this case the kinematical scattering can be considered to a certain extent as an artificial process having limited applicability. The diffraction Laue conditions appear as a result of natural limitations of the direct expansion of the solution in the resonance case.

In the framework of the developed theory, the total consideration of different geometrical diffraction schemes including sliding geometry and other surface variants proves to be possible. In this case the order of the dispersion equations does not change. This situation is related to the effective decomposition of the problem into the construction of the uniform wave field in the crystal and the determination of field amplitudes according to the boundary conditions.

Determination of the wave field as a whole without decomposition into refracted and scattered waves is the advantage of the theory. It is clear that this feature of the theory is most important for analysis of secondary diffraction processes.

We have generalized the covariant theory of dynamic X-ray diffraction to the case of a crystal with lattice deformation. In this case the displacement field is specified, starting from model representations used in Takagi-Taupin dynamic theory. In our case (in contrast to the formalism of the Takagi-Taupin equations), lattice distortions have been taken into account on various spatial scales that were different from the scale of the lattice period.

The displacement field was also a slowly varying function of coordinates. If the displacement field is considered on one spatial scale on the order of the extinction length, then the particular case of fundamental equations for the field amplitudes is obtained as a result.

In precisely this case, we have the same result as that of the Takagi-Taupin equations. By doing so we have shown possible restrictions on the applicability of the Takagi-Taupin equations to describing dynamic diffraction in crystals using various deformation models.

At the same time, the presented theory offers an opportunity for successively taking into account displacement fields of various types implemented on different spatial scales (that are larger or significantly smaller than the extinction length).

The possibility of the correct application of boundary conditions including cases of extremely asymmetric diffraction schemes in covariant theory for ideal crystals is also wholly retained for crystals with lattice distortions. Such a situation is due to the fact that the solution of the diffraction problem proper is not related to the boundary conditions; in particular the order of fundamental equations of the theory remains the same for arbitrary diffraction geometry.

Author details

Arthur Dyshekov* and Yurii Khapachev
Kabardino-Balkarian State University, Nalchik, Russia

*Address all correspondence to: dyshekov@yandex.ru

IntechOpen

References

[1] James RW. The Optical Principles of the Diffraction of X Rays. Vol. II. London: G. Bell and Sons Ltd; 1962. p. 664

[2] Pinsker ZG. X-Ray Crystal Optics. Moscow: Nauka; 1982. p. 392. [in Russian]

[3] Andreeva MA, Kuzmin RN. Mössbauer and X-ray surface optics. Moscow: Obthcenatc. Akad. Znaniii; 1996. p. 128. [in Russian]

[4] Dyshekov AA. Non-Standard Theory of X-Ray Scattering. Aktualnye Voprosy Sovremennogo Estestvoznaniya. 2009. No. 7. pp. 3-20. [in Russian]

[5] Dyshekov AA. Spatial Scaling in the Theory of X-Ray Scattering. Non-Standard Dynamic Theory. Metallofiz. Noveishie Tekhnol. 2010. Vol. 32. No 1. pp. 13-21. [in Russian]

[6] Dyshekov AA. A nonstandard dynamical theory of x-ray scattering in crystals. Journal of Surface Investigation: X-ray, Synchrotron and Neutron Techniques. 2010;4(6): 956-964

[7] Dyshekov AA, Khapachev YuP. Novye Analiticheskie Podkhody k Zadacham Rentgenodifraktsionnoi Kristallooptiki. Nalchik. KBSU. 2010. p. 75. [in Russian]

[8] Landau LD, Pitaevskii LP, Lifshitz EM. Electrodynamics of Continuous Media. 2nd ed. Oxford: Butterworth-Heinemann; 1984. p. 460

[9] Nayfeh Ali H. Perturbation Methods. New York: The Wiley Classics Library; 2000. p. 437

[10] Khapachev YP. The theory of dynamical X-ray diffraction on a superlattice. Physica Status Solidi (b). 1983;**120**(1):155-163

[11] Molodkin VB, Shpak AP, Dyshekov AA, Khapachev YuP. Dynamic scattering of X-ray and synchrotron radiation in superlattices. X-ray diffraction crystal optics of superlattices. Kiev. Akademperiodika. 2004. p. 120. [in Russian]

[12] Bushuev VA, Oreshko AP. Specular X-ray reflection from a crystal coated with an amorphous film under the conditions for strongly asymmetric noncoplanar diffraction. Physics of the Solid State. 2001;**43**(5):941-948

[13] Dyshekov AA. Generalization of a nonstandard approach in dynamic diffraction theory to the case of deformed crystal. Journal of Surface Investigation: X-Ray, Synchrotron and Neutron Techniques. 2013;7(1):56-61

[14] Dyshekov AA. Generalization of the nonstandard approach in the dynamic theory of diffraction for deformed crystals. Crystallography Reports. 2013; **58**(7):984-989

www.ingramcontent.com/pod-product-compliance
Lightning Source LLC
Chambersburg PA
CBHW081239190326
41458CB00016B/5847